The Other Side of the Coin

The Other Side of the Coin

Trust, Money and Judgement
in the Places the Banks Left Behind

Edwina A. Thompson

Wombat House Publishing

The Other Side of the Coin
Trust, Money and Judgement in the Places the Banks Left Behind

Published by Wombat House Publishing.
First published in 2026.

ISBN 978-1-7648426-3-1 (hardback with dust jacket)
ISBN 978-1-7648426-4-8 (case-laminate hardback)
ISBN 978-1-7648426-6-2 (paperback)
ISBN 978-1-7648426-5-5 (eBook)

A catalogue record for this book is available
from the British Library.
A catalogue record for this book is available
from the National Library of Australia.

For Martha de Richemont

30 March 1920 – 22 October 2024

with whom I shared a birthday, and

who left me one enduring instruction:

MOVE.

How much better to get wisdom than gold,
and good judgment than silver!

— Proverbs 16:16

Contents

AUTHOR'S NOTE

A few coordinates may help before the story begins.

For more than twenty years, I have followed a single question: how money moves — and how it can be trusted — in the places the banking system has left behind.

I have worked on that question from every side: first as a researcher, spending years among the money dealers whose networks the world calls *hawala*; then as an operator, leading teams and moving money through disasters and wars for international agencies; and as an adviser to governments, banks and law enforcement agencies trying to understand from the outside what I had watched working from the inside.

Eventually I concluded that the missing piece would have to be built. That became Amanacard: the name under which all our work travels — a system that verifies the family, worker or supplier at the far end of a payment, records what they receive and connects that handover back to the sender.

This book is not the company's story. But the company grew out of everything this book describes. When a 'we' appears in these pages, it usually means the team you will meet along the way.

There is another thread too, less visible at first. Money was never only metal, paper, plastic or code. Some forms have material value; others have almost none. What matters is not what money is made of, but what it does. Money is a language for making trust travel beyond the people we know: a promise, a name, a record, a judgement that an obligation will survive distance.

The strongest trust is still built in relationship, which is why artificial intelligence now matters to this story. I did not set out to write about AI, but the question it raises — whether systems that can speak in the register of intimacy, remember, classify and act at scale can be made answerable to human life — was already inside the work.

Introduction

TRUST, TWENTY-FIVE YEARS ON

I heard about the towers by telephone.

In September 2001 I was working for Kevin Comolli at Accel Partners, one of the world's biggest venture firms, about as deep inside the citadel of Western capital as a twenty-two-year-old English literature graduate could get. Kevin was in the air, London to San Francisco, when America shut its sky. The pilot let him use the phone — a rare privilege in those days — and so I learned what was happening from a man suspended, absurdly calm, above a world descending into chaos: airports sealing beneath him, flights scattering to whatever country would take them, his own plane bending north towards Edmonton, Canada, where the passengers would be shepherded into hotels to share rooms with strangers. What I remember most is the confusion — time and space compressing and stretching at once, and how difficult it was to make sense of what we were hearing. Then, like everyone else in London, we watched the horror repeat on television, and the rest is history.

I have often thought about where I happened to be standing that morning. I was observing one end of the financial world from a few feet away, among people who could move enormous sums with a telephone call. In the twenty-five years since, I have spent much of my time observing the other end: among money dealers and trust-based networks that kept entire economies moving where banks could not or would not. I never quite belonged to either world. I was simply close enough to see how readily one was trusted by the West, and how quickly the other was condemned.

For the money dealers, the blame arrived within weeks, from the highest podium on earth. In the Rose Garden,

President Bush declared a "strike on the financial foundation of the global terror network". Soon afterwards, as Al Taqwa and Al Barakaat were shut down, he described networks that presented themselves as "legitimate businesses", while operating "all at the service of mass murderers". A victory, the White House said, in America's fight against evil.

I watched the coverage from afar and came away curious rather than convinced: *hawala* — an ancient means of transferring value through trusted brokers rather than banks — had just been indicted from a lawn, and I could find almost nobody who could tell me how it actually worked.

Several years later, during my doctoral research, I was tracing gun-trafficking networks in the Highlands of Papua New Guinea and the money that moved behind them — a trail that led, as such trails regularly do, into offshore financial centres: paper, shell companies, silence. The further I followed the money, the further I moved from people, which is to say away from the interesting part — what actually makes things move and work.

Afghanistan promised the opposite: a financial system you could understand only by drinking tea with the human beings who ran it. So I shifted my focus and went to the money men themselves — on an Australian postgraduate scholarship that gave me enough to live on and the freedom to follow the question. Not because anyone was seriously asking how *hawala* worked. By then, governments, banks and the media had largely decided that they already knew.

That question became a trade. For the next twenty-five years, when money had to move somewhere the banks would not go — famine, typhoon, siege — it kept being my phone that rang.

I found the mystery was not peculiar to *hawala*. Correspondent banking also moved value through chains of relationships, with accounts settled between intermediaries and each institution depending on another to honour its side of the transaction. Yet one was called global finance; the other, a foreign scourge. I wanted to understand why.

There is a family joke in all of this. I am hopeless at mental arithmetic. I have spent most of my life among traders, bankers and money dealers who can calculate a profit, a commission and three currency conversions before I have found the calculator on my phone. That hopelessness with numbers may be why I became so devoted to records and reconciliations — and why I struggle to trust anything I cannot check, most of all my own calculations.

That is where the proverb at the front of this book becomes practical: wisdom better than gold, judgement better than silver. It can sound, at first, like a moral ranking of spiritual things above material ones. But gold and silver only work as money because people recognise them, protect them, exchange them and account for them. Paper money, bank balances and digital records ask for an even greater act of trust. They depend on judgement: who is believed to determine value, whose claim is recorded, and who must answer when the record is wrong. After 9/11, the West largely treated exclusion as the price of trust.

What the West built instead of trust

The financial response to 9/11 became an architecture

of suspicion so vast that it is now almost invisible to the people who live comfortably inside it: sanctions lists layered on sanctions lists, compliance departments larger than the aid programmes they police, banks that first profited handsomely, for decades, from the custom of despots — then were fined into reflexive fear, and finally praised for withdrawing from entire nations. The stated purpose was to protect the system from abuse. The observable effect, out on the frontier, has been the steady exclusion of everyone who cannot afford to prove their innocence: the charity that cannot bank, the nurse whose remittance bounces, the family in a blocked corridor, the honest merchant whose town has the wrong postcode, or none at all. We built a fortress against the few and locked out the many — and the few, it must be said, have generally found the service entrance.

I began with an optimistic mistake. I believed these failures persisted because people did not know: that if the evidence could be made clear enough, and a practical alternative built, reasonable institutions would choose differently. I built a working life on that belief.

The chapters that follow are the record of how that belief is still being tested.

The book behind this book

In 2011, Oxford University Press published my earlier doctoral thesis, word for word, as *Trust is the Coin of the Realm: Lessons from the Money Men in Afghanistan*. I was embarrassed by it at the time — I never felt qualified to have written a doctorate, let alone a book — and for years it sat quietly on university shelves, priced beyond most ordinary readers as an academic textbook.

The book had a strange afterlife, though. A friend who ranks among Britain's most senior counter-financial-crime experts told me recently, speaking of the small international circle of expert witnesses who give evidence in prosecutions against money launderers and terrorist financiers: "You are mentioned in everybody's CV — including mine." For close to a decade, I have helped train that community. Beneath all the academic apparatus, the book has apparently continued to say something useful.

So I am bringing it back as an archive edition for those who need the ledgers, histories and footnotes.

This is the book you can read on two flights. It is about what happened when I stopped studying the problem and began operating inside it — in what Lord Jim O'Neill, who coined BRICs, called the 'next frontier' of fintech.

An education in grey

I was a black-and-white person when I started, convinced — as the young and fortunate often are — that problems persist because nobody has explained them clearly enough. The world corrected me one introduction at a time, over years of work as a volunteer teacher, aid worker, researcher and consultant.

First: Borneo, before any war zone. I had paid for the trip in the only way I could then imagine: by turning a bundle of school poems into a small book, *Knee Deep in Scribble*, selling it through the school community and having my headmaster match what I raised. In the interior, a religious cult was bringing false hope to a community that had little else. It was very cruel, and nobody involved thought of themselves as cruel. That distinction has followed me ever since.

Second: Dragan, back in London — a jet-engine engineer stacking supermarket shelves. I could not understand why someone would leave a place he loved and a profession he had mastered. He explained the violence of the Yugoslav wars, which had made staying impossible — and made me think, for the first time, about what refugees were forced to leave behind.

And later, in Somalia, I met hunger: an old man carrying a handwritten letter I could not answer.

In Darfur I met shame: a woman raped while collecting firewood after a gap in donor funding closed the programme that had made the journey safer. Her shame was imposed by her community. Mine came from belonging to a system that could describe her risk perfectly yet still leave her exposed.

In Afghanistan I met fear: bargaining with a warlord over how much rice he would take from convoys intended for pregnant women.

In Haiti I met panic: watching large distributions and military logistics push the elderly and orphaned to the edges of the queue.

And in a market in northern Syria, in the middle of war, I met freedom: a merchant standing behind a mountain of potatoes, still trading, employing people and insisting on having a future.

His greatest problem, he told me, was not the war. It was competing with imported aid — poor-quality food and hygiene kits that people did not want, resold cheaply in the market and undercutting the goods they had actually chosen to buy.

He understood something the aid system repeatedly forgets: a crisis does not turn people from economic actors into passive beneficiaries.

Cash was the natural disruptor. Where there was cash, trade returned; and where there was trade, there were money men.

There were many more such encounters across Asia — Cambodia during election unrest, Laos and Vietnam in the nineties, Sri Lanka as Tamil civilians were driven into the Vanni's 'safe zone' that did not save them, the Philippines after the Super Typhoon — and on through the Middle East, Africa and the Americas, as far as the offshore ledgers of Belize and the Caymans in the wake of the Panama Papers — eleven million files taken from a single Panamanian law firm, exposing everything from lawful loopholes to plain crime dressed in paperwork. On those same shores ran a second shadow economy, this one in the flesh: packages of cocaine slipping off boats by night and washing up on the beaches, addiction and crime quietly taking communities apart. Two hidden economies shared one coastline — one passing through on paper for people far away, the other landing on the people who lived there.

I set these introductions down with hesitation, because they were not my tragedies. The people in them were not lessons; they were people, having the worst days of their lives while I was, among other things, having an education. The debt runs in one direction. This book is one small instalment on that debt.

The world, it turns out, is grey wherever money moves. The work is not to eliminate the grey — it is to make good things possible within it.

One of the more personal corrections began when I stopped advising other people's systems and started trying to build one of my own. As a first-time founder raising capital for a technology company, I made a slide of five women in business I thought might help me. The people who actually showed up were mostly men — another lesson in how rarely trust respects the categories we assign it. That story belongs to the women in the market, in Chapter Two.

The speed of a label

When I wrote the thesis, globalisation was still a debate. Now it is the weather, and some of its fiercest battles are over information. My field — how money moves — is one of them, and occasionally the front line runs straight through your own name.

By then, the start-up I had founded was known as Amanacard, our registered brand — the name by which people knew our work across very different parts of the world.

In November 2024 the United States sanctioned an organisation on the other side of the world whose long name was shortened in headlines and screening databases to one word: *Amana*. Where I was, in Melbourne, it was already the following day. Before I had properly understood what had happened, my phone was lighting up with messages from

Syria and Yemen. People who depended on us were being sent clips through their own networks and asking whether Amanacard — whether we — had just been sanctioned.

We had chosen the name with care. In Arabic, *amana* is the thing given into your safekeeping, and the obligation to guard it; *amin*, from the same root, is the person trustworthy enough to hold it. In Hebrew, the root gives *amen* — confirmation — and *amanah*, a covenant made firm; faith itself, *iman* in one language and *emunah* in the other, grows from the same three letters. Every branch comes back to trust: what can be relied upon, and who. We chose it as a standard to be held to; whether we have lived up to it is for others to say.

At this point, a word intended to convey trust had been swallowed by the machinery of suspicion.

I began writing. The US Treasury replied within hours and arranged a call, which under the circumstances I was very happy to take at nearly two in the morning. By the following day, Treasury officials had confirmed in writing that Amanacard was entirely separate from the sanctioned organisation, that banks were free to deal with us, and that Treasury's compliance hotline would tell them the same. They then introduced me personally to their Canadian counterparts. Ottawa took a little longer, but produced a signed letter I could share publicly.

The American designation had not been the first. Canada had listed *Amana* in June and Britain in October. Two months after the US announcement, the executive order behind the entire American West Bank sanctions programme was revoked, and OFAC, the US Treasury's sanctions office, removed every designation made under it, *Amana* included.

That was a consequence of the programme ending, not a finding about *Amana* itself. The Canadian and British listings remained.

I had also contacted the relevant British authority, including the official who should have been able to resolve the confusion created by the listing. The remedy was not complicated: confirm that Amanacard was entirely separate, and stop abbreviating the sanctioned organisation's full name to the single word *Amana*. Some eighteen months later, the answer is presumably still receiving careful consideration.

Swift as the American and Canadian responses were, the deeper damage had already been done. The bare name had been absorbed into the large screening and data systems on which banks increasingly rely. A human being could read the clarification and understand the distinction. A database merely retained another association. The suspicion had become data; the correction was still only a letter.

Since then, we have faced a steady stream of additional questions and de-risking decisions. I cannot prove that each one began with this episode; correlation is not causation, particularly in banking, where the reasons for a decision are withheld from the person affected by it. But the timing is difficult to ignore. I find myself repeatedly addressing variations on the same concern — and mostly not even the useful questions, such as how our controls work or where the money actually goes.

I answer them, because that is what one does when access to the financial system depends on correcting a machine's first impression and educating humans, sometimes repeatedly within the same bank. But setting the record straight is much harder than putting something into it in the first place. The allegation travels automatically. The correction has to be carried from institution to institution, explained by a person and believed afresh each time.

I keep the US email and the Canadian letter together, with a space reserved for the British reply: a small museum of how three systems responded when a word travelled faster than its meaning — and how difficult it is to call it back once the databases have taken note.

A word is never only a word once a machine can act on it. It becomes an instruction, a warning, a refusal, the reason a payment stalls or an account closes. Language is the medium through which institutions identify us, classify us, misclassify us and pass judgement onward. In AI, a confident falsehood is often called a hallucination. In human life, it is a blocked payment, a duplicated profile, a person who is not recognised. That is why names matter in this book. So do translations, ledgers, lists, categories, promises and corrections. They are not decoration around the system. They are often the system itself.

What you are holding

This book is the sequel the archive edition never expected. Each chapter takes one idea from the 2011 book and follows it into the present to see what it did when it grew up.

The evidence is not what you might expect of a book about money. It includes a bikini wax, a burned notebook, a bazaar whose books held New York's accounts, bakers hired to build trust under bombardment, bin bags full of hospital cards and two gold coins under armed guard, fireflies at an open-air cinema in a flattened city, a seatbelt unbuckled at speed in Baghdad, three thousand names on a kitchen table, a question that silenced a room convened by the World Bank, a pile of cabbages in the dirt that put corporate power to the test, a dream about my father, and one letter I have never been able to answer.

If you work in a bank, start with the theatre. If you work in aid, start with the banks that left. If you write sustainability reports, start behind the kitchens. If you have been away from home long enough for it to become foreign, start with the book that must not burn. If you are young and about to go somewhere difficult with a notebook, start with the women in the market — I am cheering for you.

The institutions in these pages often had the luxury of distance. The people living with their decisions did not. The rest of this book is about what it costs to misread people, places and systems — and what it takes, on the ground, to understand them well enough to build something better. It was written from the space between them.

1

THE COIN FLIP

In the weeks after 9/11, the West made one of the fastest financial judgements in modern history: heads the banks, tails the bazaar. Heads won — it usually does when the coin belongs to the house. This chapter is about the coin itself: what I found when I picked it up and read both faces.

At first sight, Abdul Haq seemed as far from global capitalism as it is possible to be. Wearing a traditional *shalwar kameez* — a long, loose tunic over baggy trousers — he sat in a small shop in Kabul's money bazaar, one of hundreds of dealers conducting business across several crowded floors concealed behind narrow alleyways. The entrances were so easily missed that Western visitors could walk straight past without suspecting that a financial system was operating within. There was no marble, no glass tower, no glowing wall of screens — just a trade conducted at speaking volume, rates travelling by voice, money counted in the open, and tea arriving for anyone who sat down long enough to matter.

I did not walk past, but I hardly arrived equipped: a young woman with a notebook, no tape recorder, and — in those first days — an interpreter so unsuited to the place that his presence was itself a lesson, one I will explain in the next chapter. The bazaar absorbed us both with the patience of an institution long accustomed to outsiders arriving without understanding what they were looking at.

On the shelves of Abdul Haq's shop stood a row of lever-arch files, their spines scrawled in Pashto. Nothing about them announced their contents; that was rather the point. I remember the moment he let me open one — a permission earned in tea and returning, though I did not understand its significance until much later. Inside, a clear hand tracked the running balances of his clients: international organisations headquartered in New York. Private companies in the Gulf. Families in three countries, their remittances and school fees and debts moving through the same pages. The man the newspapers had spent four years calling a terrorist banker was keeping the books for the people sent to rebuild Afghanistan — and keeping them better, I would come to learn, than most of his accusers kept their own.

The mechanism behind those balances fits in a few sentences. A man in Toronto hands cash to a merchant and names a recipient; the merchant calls his counterpart in Kandahar, and the counterpart pays out before the tea has cooled. Nothing crosses the border at first except language: a name, an amount, a promise. What crosses later is the settlement — squared between the two merchants against physical trade, remittances and a running ledger of favours, exactly as the great books recorded each evening. The system is not the absence of banking. It is banking with the collateral made of reputation: one person's word becoming value because another person is willing to act on it.

Everything I have built since has been, one way or another, an attempt to make what was in those files legible to the world that was calling Abdul Haq a terrorist — without destroying the trust that made the files true.

I had come to Afghanistan in 2005 with the question of the hour after September 11: how did terrorists move money to facilitate their attacks? Except almost nobody in Washington or London was still genuinely asking. They had largely decided they already knew; the coin was down, and the answer was *hawala*.

I went to ask anyway. For nine straight months I worked through the money markets of Kabul, Herat, Mazar-e Sharif, Jalalabad, Gardez and Faizabad, then across to Peshawar's Chowk Yadgar and down to Dubai. This was not a succession of neat interviews. I returned to the same shops again and again, sitting for hours over months, drinking tea, watching transactions, checking explanations against what happened next, and slowly earning the right to ask better questions. I worked with fifty-four dealers, all men with fourteen to twenty-five years in the trade.

By the end, the question I had arrived with had come apart in my hands.

The received wisdom of the post-9/11 era was built in a hurry. Weeks after the attacks, a *Time* magazine headline declared *hawala* "A Banking System Built for Terrorism". The phrase hardened into fact by repetition — quoted in briefings, recycled in threat assessments, embedded in policy. Yet when the 9/11 Commission finally reported, it found that the plotters had made extensive use of the formal international financial system. Mohamed Atta and Marwan al-Shehhi held a joint account at SunTrust in Florida, receiving US$109,500 in four bank-to-bank transfers from Dubai through Citibank's correspondent network. The irony was profound, and almost nobody sat with it: the plot's money had travelled largely through banks, while the system being cast as inherently criminal was the one al-Qaeda's core relied on in Afghanistan, where banking was rudimentary and unreliable. The rail did not reveal the purpose.

What I found in the shops was a coin with two faces, held

by the same hand. The same dealer who settled accounts for a United Nations agency moved the savings of a refugee family in Karachi. In Faizabad, the capital of Badakhshan province — a town so remote that the far end of its province narrows into the Wakhan corridor, that strange finger of Afghanistan pointing between the Pamirs towards China's Xinjiang, the last place on earth you would think to look for serious money — one dealer could deliver US$450,000 to a UN programme within two hours. Always. During the currency changeover of 2002, when the entire country's banknotes had to be replaced in months, it was the money dealers — not the banks, which had effectively ceased to exist — who exchanged the old notes in the first two weeks. Aid agencies moved hundreds of millions of dollars through these men: wheat programmes in Badakhshan, demobilisation payments to ex-combatants, election logistics. Nine out of ten agencies I surveyed then, and every charity I have asked since, reduced the decision to two criteria.

He delivers.

He is cheapest.

The pronoun is deliberate: all fifty-four *hawala* agents in my survey were men. The exception would come years later, during the war in Syria — Samira, a woman running a substantial warehouse and logistics business who took some of the most treacherous routes in the country to the hospitals we were helping to keep running. My field team found my excitement about her slightly strange. I still think she belongs to an epic of her own.

Back in Afghanistan, the same networks had another face. In Helmand and Kandahar, merchants told me plainly that eight or nine out of every ten of their colleagues handled drug money. My estimates of those flows — research funded partly by the UN Office on Drugs and Crime and published by the World Bank — were the first attempt to quantify the overlap between the money trade and the drug trade. One senior UN logistician, asked how his agency squared its payments with everything else moving through the market, gave me the most honest sentence of the whole fieldwork: "Of course we are all aware that we are directly laundering drug money. But it is not our concern... It is just too hard to think through what we are actually doing."

Two faces. One coin. The policy world flipped it rather than read it — and which face came up depended less on evidence than on whom the room was prepared to trust. That observation became the book this one grows out of. I did not know then that I would spend the next twenty years holding the coin myself.

The same coin, twenty years on

I should confess what I believed when I started, because it is the first belief this book will retire. I thought I was documenting something ancient in its final years — a folk system the banks would soon replace, worth recording before it disappeared.

I had it entirely backwards.

Twenty years later, the system I had expected to disappear had adapted without breaking stride. It was the banks that disappeared from whole markets — or proved themselves to be the institutions people feared would simply keep their money. I have lost count of the merchants I have watched plead with banks in Iraq, Turkey and Sudan to release funds sitting in accounts that were indisputably theirs. Sometimes I could help. Sometimes I could only stand beside them while the bank declined to explain.

I encountered a milder but revealing version on the British high street. Before travelling to Syria, where cash was the only form of money I could use in the places we were supporting, I would try to withdraw US$10,000 or US$20,000 from our company bank account. My Syrian colleagues could not understand why this was difficult. Neither, initially, could I. The cash had to be ordered, transported to a central London branch and collected more than a week later a small reminder that the supposedly modern system was not always very good at producing money when somebody actually needed it.

The bazaar has smartphones now — even iPhones, a small revelation that belongs to Chapter Six. The lever-arch files have become spreadsheets, some of the tea is taken over FaceTime, and value can now travel as stablecoins — cryptocurrencies designed to hold their value against a conventional currency such as the US dollar. Yet the merchant keeping a displaced district supplied with medicine is still, in a compliance file, easily mistaken for a risk. The network that moved famine relief faster than the aid system is still, on a screening result, an alert.

The banks did not learn to read what they were seeing. They withdrew instead — corridor after corridor, from Mogadishu to Kabul, in the great retreat regulators politely call de-risking.

For years I observed both faces of the coin. Then I found myself working along its edge, where the institutions that controlled the money met the people who needed it to arrive. I can place the change.

In a London meeting room sat Barclays and the regional director of one of its clients, a humanitarian organisation trying to move US and European funding to hospitals inside opposition-held Syria. The money had been blocked for seven months. Doctors and nurses had continued working without pay. The bank could not see enough to say yes; the charity could not make the people and transactions visible enough to be believed. Between them sat the oldest problem in this book.

What unlocked the funds was a proposition: release the money, and we will track every dollar to its recipient, whom we will verify ourselves — each named person confirming the next, with a system preserving the chain all the way down.

Barclays agreed.

That meeting marked the beginning of what became Amanacard, and also of my unusually intimate relationship with WhatsApp. My phone became a continuous record showing that cash could be accounted for in places no auditor would ever stand.

At the annual expert witness course I help teach in the UK, I had always spoken from the researcher's distance. Now I looked uncomfortably like the figure the course calls a 'Controller' — the person at the centre of the network who knows every face.

So more recently I chanced it. With the participants' permission, I put a selection from my live WhatsApp feed on the screen, on the understanding that no recording would be made. Most of the stream was in Arabic, so the room had little help from the words. I asked the course attendees — investigators and financial-crime specialists from the FBI, Australian Federal Police, the UK's HM Revenue & Customs and elsewhere — what they thought they were looking at.

They agreed unanimously that it was criminal activity.

Then I supplied the context and introduced the faces one by one: a hospital administrator in a war zone; a field team member; a merchant mid-delivery. The exercise taught them something about evidence. It taught me how the world I worked in looked from the other side of the coin.

Through those messages, the Syrian money magicians began to emerge from behind their tricks. During one hospital payment, Abu Wajih sat among ashtrays and unmarked black bags while administrators counted the cash he had delivered. A blood-collection team had set up beside them.

The photograph could hardly have looked more incriminating. Yet the bags held a hospital payment, the merchant had delivered it, and the needle in his arm was there because, rather than waste the wait, he had decided to donate blood.

Supplying the context became the work of my second twenty years, inverting the work of the first. As a researcher, I translated the money men to institutions. As an operator, I built that translation into infrastructure: verified recipients, named local agents recording what they observed in their own languages, a human signature on every judgement, and an audit trail from the source of funds to the end destination.

Not because the technology was clever, but because it allowed a regulated institution to see the transaction in context — and remain answerable for the decision it made.

People often decide whose account carries weight before they examine the evidence. The bazaar taught me that before I had language for it. The question did not remain in the bazaar.

Ten years into the intervention, I was asked to assess what ten years of UK aid to Afghanistan had achieved. The Afghans I interviewed — professionals, business people, farmers, aid workers — kept returning to the idea and extent of time itself: we wanted quick wins where the need was generational. No meeting in Afghanistan runs long without a proverb, and the one that kept surfacing carried an old ethos of patience: 'You have the watches; we have the time.'

Afghans were also tired of being surveyed. Too many people had arrived with questions, extracted their answers and disappeared. I told them that this time their reflections would be carried to the parliamentary committee examining Afghanistan, and framed the exercise as a message to the British Prime Minister. That promise persuaded many of them to try once more. They gave me thoughtful, balanced and often deeply uncomfortable accounts of the decade behind them.

The report acknowledged what aid had achieved, but gathered its failures around six constraints: conflict and criminality, corruption, capacity, competing interests, complexity and, finally, common sense.

The night before its launch, we were told that one of Britain's largest aid agencies — one of the organisations behind the research — had been instructed to remove its logo. For several alarming hours, we thought the entire report might have to be withdrawn. The other agencies held their ground. The event remained in the diary. We reprinted every cover overnight.

By morning, it was the same report with one fewer logo.

I presented its findings at the House of Lords. When the aid department's senior representative rose to respond, he faced me rather than the room. The papers in his hand shook. He delivered his entire response at a tempo wildly out of proportion to the setting. It felt less like a discussion of research than a dressing-down for a schoolgirl caught once too often smoking behind the toilet block. From first word to last, he attacked my credibility, judgement and methodology. He did not address what the Afghans had said. He addressed the person who had delivered their message.

I think the final C was not especially well received. Conflict and corruption could still be located in Afghanistan. A failure of common sense, however, pointed back at the people designing, funding and defending the aid. The timing cannot have helped. The department was, as I understood it, making its case within government for the aid budget, and this was hardly a convenient moment for such an uncomfortable message from Afghans.

Standing there, I realised that I was living out one of the report's own findings. Afghans had said that the international community had "many tongues, but not many ears". Reports accumulated. Lessons were identified. Inconvenient voices were discredited and added to the pile.

I had become one of them.

The communities during my doctoral research had already taught me to keep a longer clock. They spoke of events from decades earlier as though they had happened yesterday, and assumed that anyone serious would do the same. Perhaps that is the difference between the two faces of the coin. One side flips. The other reads.

2

THE WOMEN IN THE MARKET

There were no women in the money bazaar. Not behind the counters, not among the customers pressing in with bundles of afghanis and dollars and Pakistani rupees, not in the back rooms where the day's balances were settled. When I walked into that world in the early 2000s — a young Australian woman with a notebook — I was not breaking a glass ceiling. I was simply outside the categories the market had. And that, it turned out, was the point.

My first day was a small disaster, because I was a rookie and made rookie mistakes. I had chosen my interpreter for his excellent English, on the recommendation of the BBC journalists I had met during one of my sparing visits to L'Atmosphère, the improbable poolside enclave where expatriate Kabul gathered. It was the first and last time I hired on language skills alone. Two things put us both in danger, neither of them through any fault of his own: his private life in that strict environment and the fact that he knew nobody in the bazaar — which, in that economy of introductions, meant that neither, suddenly, did I.

I was not ready for the ceremony of it all either — the hospitality that descends on a foreigner alone, the tea that must be drunk, the order in which respect is paid. That first day it took the form of the biggest pile of skewered meat I had ever seen, glistening with oil — every other skewer, I discovered, was simply rows of fat — set down in a money dealer's office as what felt like hundreds of staring men formed a circle around me. I asked them to start. They refused: all of it was for me. Wow, did I feel sick later. But what a gracious gesture it was — my first practical lesson in

Pashtunwali, the tribal code that obliges a host to honour the foreigner, even an enemy. I had arrived a fussy eater; the bazaar cured me of that for life.

The lesson I took from that first stumbling day shaped every decision afterwards: in a world that runs on vouching, you sit beside the person who introduced you, and you stay there until the room has decided what you are.

And the room's first decision, for the record, was that I might be a spy. What dissolved the suspicion was not an argument. I put it down to the eyes. Unlike the Americans and the military — the only other people out beyond the wire — I never wore sunglasses. I wanted to eyeball everyone, and be eyeballed back, because there is something universally human in it: you can see through people, and be seen through, in ways language never manages.

Women held my gaze from inside their coverings; only the full burqa defeated it — a grid where a face should be. At first all that pleated blue fabric seemed impossible to miss. Yet against the pale Afghan sky, the same colour somehow made the woman inside it recede, until I had the shock of catching myself no longer registering women in the street. That was the disturbing part: not only that the burqa concealed a face, but how quickly the eye could learn to let an entire person fade into the background.

It was not until my first visit to an all-women's market, where the burqas were lifted, that I saw the full spectrum of the other half of Afghanistan — women my age, women my mother's age, little babies tucked under their shirts.

Years later, I understood more fully why eyes had mattered to me. My beloved uncle suffered a sudden brain aneurysm and fell into a coma. By an extraordinary accident of timing, I was there with his three children when they were forced to think through what would happen when the machine keeping him alive was switched off — and which parts of their father, if still viable, might be given to strangers.

They were trying to make decisions of permanent consequence while still absorbing the fact that the man they loved was not coming back. One possible donation after another could somehow be approached through the language of medicine and usefulness. Then the question of his eyes stopped them.

I understood why. His heart had sustained him; his lungs had carried his breath. But his eyes were how we knew him. Through them he had recognised his children across a room, registered a joke before he laughed and looked back at all of us for as long as we could remember. The thought that even part of those eyes might be given to someone else felt different from the donation of anything else. It came unbearably close to the person himself.

That experience sharpened something I had been learning for years: how much of a person is communicated before any formal credential appears. I have always been uneasy about iris-scanning technology, and have avoided it in everything we built. Eyes feel like the most personal thing a person can own or offer, and I have spent a career being offered them.

In the bazaar, that instinct became practical. The gaze did what no credential could: it opened doors, and returning through them slowly built trust.

The gaze also gave me access to the part of the economy conducted behind the household wall. In some of the most conservative places on earth, women governed what the family ate, wore and spent; which debt was repaid; and whether a daughter remained in school. I could sit with them. When I stayed with families, the women and girls slept around me on the floor — no privacy, but an intimacy no formal interview could have produced. That access made the whole journey visible — from a name and number in a merchant's book to the life the remittance funded: a school fee, a debt cleared, another year of education bought.

The salon

One of the more memorable episodes of my early days in Kabul had nothing to do with terrorism or money. It was my first Brazilian bikini wax. Kabul was not, in retrospect, the ideal place to begin.

The beautician was a student, surrounded by other students who watched my initiation with considerable interest. She approached the task with great confidence and, as far as I could tell, one enormous sheet of wax. I have since learned that this is not the recommended technique.

Twenty years later, I was working in Afghanistan again as the Taliban closed the country's beauty salons. I had gone for an entirely ordinary reason: to feel normal among other women in an unfamiliar place. What I had not understood was what that room made possible for them. Behind its closed door, women could earn money, remove their veils, laugh, gossip, complain, celebrate weddings, share grief and exist together without the constant presence of men. It took that long and a decree for me to understand the privilege I had been given — not the wax, which I could happily have done without, but admission to the room.

I would not pretend that wearing a burqa once taught me what it meant to live inside one. It taught me something narrower: how completely it could remove a woman from public view.

In Jalalabad, that disappearance saved my life.

The escape

A week before a TEDx event in London, one of the speakers withdrew. Asked whether he knew anyone interesting enough to fill the slot, my big brother very kindly volunteered his sister. The rules were unforgiving: no notes and no more than seventeen minutes.

My sister's friend and colleague EJ, a gifted communicator, patiently helped me learn the words; Ben, my close friend from school and a trained actor, discovered that I did not know where my diaphragm was. He taught me how to find it and, if nerves took over, told me to lie flat on the floor.

Minutes before going onstage, I was therefore flat on my back in the preparation room, wearing a burqa and trying to breathe. People I trusted had persuaded me that telling the story might help others understand what had happened. The burqa was stagecraft.

The first time I had pulled one over my head, I had been trying to disappear.

In May 2005, the access I had worked so hard to build offered no protection. A front-page *Newsweek* report — unverified and later retracted — that interrogators at

Guantánamo had flushed a Qur'an down a toilet travelled across Afghanistan faster than any correction could follow. Jalalabad erupted.

The building where I had been interviewing a drug dealer came under attack. A Norwegian diplomat who knew where I was tried to warn me by email. The dial-up connection had not finished logging me in when AK-47 fire shattered the line of computers along the wall. As the building was looted and set alight, I escaped over the back wall. Somewhere between the gunfire and the climb, I lost my headscarf.

It was the first time I truly thought I would die, and I met my own adrenaline properly — in both directions at once. It moved me when movement counted. It also made my hands shake too violently to work the buttons on my mobile and call my family. The phone was soon a casualty anyway, taking all my hard-earned contacts with it.

The house where I then hid was not safe for long. Young men began gathering to discuss my ransom value. An elderly *mullah* sat in the corner of the room, fingering his prayer beads and staring at me — not with the curious, appraising gaze I had encountered in the bazaar, but with contempt.

Overhead, coalition helicopters beat the air with a deep, pulsing thud. I remember doing the arithmetic clearly: if the mob did not get me, the men further up the kidnapping chain might; and if neither did, friendly fire might finish the job by accident.

I knew a waiting game was not in my favour, so I made my own plan. I managed to send a message through a little girl to two local friends, who smuggled a burqa into the compound. In a moment of distraction, I pulled it on and ran.

The same blue grid that had once made women disappear from my sight now allowed me to disappear from the mob. Looking through it, I navigated my way past burning cars to the place where I had been told to stand. I jumped into a car and we made a terrifying drive to the airport. I ran across the tarmac and begged my way onto a former Soviet military Mi-17 helicopter leaving the city.

I took refuge at the Norwegian diplomat's residence in Peshawar. It was probably modest in ordinary circumstances, but the garden and a refrigerator full of fresh fruit felt like paradise. Then I went back to work, because the research was not finished.

Two acts of kindness remain clearer than much of the danger. The first belonged to the little girl who connected me to the outside world from inside a hostile place. I never learned her name.

The second belonged to the person I relentlessly persuaded to return to the building and search the room where I had earlier been told to stay and, at one point, locked inside. By then, its contents had melted. In one corner he found my notebook, burned along every edge but still legible through the middle.

The fieldwork had survived.

Field data is often cited as though it simply appeared. Mine had to be recovered from a room that had burned around it. I have been wary ever since of over-confident conclusions that acquire authority in conference rooms without ever being tested in the places they describe.

Two women, one market

Twenty years after the room decided what I was, the Afghan operation I founded is run by a woman. Maryam coordinates, in the hardest market we serve, everything I once studied — the dealers, the settlements, the verifications — under a government that has erased women from public life, and she absorbs treatment that would end careers if it happened in London: not only from the men you would expect, but from inside institutions that print their values on their letterheads. Her best support includes the men on our own team who choose, every day, to make her work a little easier. I have thought hard, more than once, about closing that office. We have not — in part because her salary holds her whole family: her father is dead, and her sisters are not permitted to work. That is what a job means there, and what closing an office would mean.

In 2016, when we named the service, we chose *amana*, a word carrying trust across Jewish, Christian and Muslim traditions. When we tested the name in Venezuela years later, it landed well there too: it carried, we were told, feminine and nurturing connotations.

I have stopped treating that as a coincidence. In every fragile economy I have worked in, across five continents, the last mile of trust is disproportionately held by women — the household treasurers, the ration-stretchers, the ones who know exactly what arrived and what did not. She is the true last mile of every payment system on earth — and most of them have never tried to reach her. When we finally built our own, that finding governed the design: alongside individual accounts we issued a household identification number, so a family could share a card securely. The gatekeepers said yes, and the women were no longer locked out. It remains the least glamorous and most effective inclusion feature we ever built into the platform.

The inversion

Here is what I could not have predicted at twenty-something, filling notebooks in a bazaar that had no category for me: being a woman helped me gain access in the field but would later deny me authority in the boardroom. The money men of Kabul decided within weeks that I could be trusted with the truth of their trade. The men I met in global finance took years longer, and some never got there. I still sit in meetings where my own research is explained to me without their realising. I have seen newer arrivals — often young men with family roots in the countries where I had worked, a biography that understandably reads as authority — announce themselves as the first to open corridors I had been operating for a decade, and watched rooms believe them because their story fitted better than mine did. A Western woman: what would she know?

My doctorate was about exactly this, and it still took two decades for the lesson to sink in — the same lesson I shared at the expert witness course: we weigh the person before we weigh the evidence. The bazaar decided on observation — she came back, she understood the ledgers, the elders vouch for her. The boardroom decided on resemblance — who best fitted its idea of authority. Of the two systems, the bazaar's was the more rigorous. I offer that finding without bitterness, as a matter of record, and I have stopped apologising for it: the discounting of my testimony is itself a data point in the thesis of this book.

The five-women slide

For years, I believed the correction would come from women — that female identity would naturally produce female alliance. I was wrong, and the way I was wrong taught me more than being right would have.

In February 2019, I spoke at a wealth-management event at Spencer House in London alongside Brenda Trenowden and my namesake Edwina Dunn — women who had changed British boardrooms and taught business to take data and loyalty seriously. Women ahead of me, in every sense.

At an event about women supporting women, I made a confession. When I first set out to build Amanacard, I created a slide showing the five women in business who most inspired me. I wanted them on my board.

Although I got none of them, what arrived instead was a succession of men who showed up when it counted. After September 11, Kevin wrote me a personal cheque that funded nearly a year of refugee-protection work at Amnesty International, where Dr Graham Thom taught me my trade. Two decades later, former colleagues at Goldman Sachs introduced me to Lord Jim O'Neill, who heard the Amanacard pitch over lunch and provided the seed investment, describing it as the next frontier of fintech.

Huub Spierings was our first angel, in every sense of the

word. He was a close family friend: generous, loyal and equipped with the sharp commercial mind of a man who had risen from grain trader to running Cargill's European business. He cared about what happened to us, not only what happened to his investment.

Huub also saw the central danger before I did. If governments failed to adopt the solution once we had proved it, he warned, private capital could not sustain it indefinitely; we were trying to solve a public-good problem. That left an awkward gap: private capital could prove the thing, but the public institutions needed to sustain it could take years to act, if they acted at all. I heard sensible investor caution. He was describing much of what followed. Huub has since died, and I miss both his enormous heart and the mind that understood the structure of the problem so early.

The uncomfortable lesson of the five-women slide is not that I chose the wrong five women. It is that I had made the same kind of shortcut I had spent my career examining. Institutions saw *hawaladar* and inferred danger. I saw woman in business and inferred ally. The consequences were not comparable, but the habit of mind was: I had asked a category to predict a person's conduct.

It could not.

That does not make women less important to one another, or men natural allies. Solidarity is something people demonstrate, not something identity guarantees; Maryam's working day is the proof. The useful distinction is not between women and men. It is between those who show up and those who do not.

The same correction followed me into how aid systems

regard the people placed at the bottom of their diagrams. When we launched the gold Amanacard for families, alongside sky-blue cards for staff and royal-blue cards for suppliers, some aid organisations objected. Gold was absurd for people who could not afford it, they said. It might make them look rich.

My answer was that hardship should not condemn anyone to the cheapest object, the worst interface or an experience designed to remind them of their poverty. Dignity was not an inappropriate luxury. It was part of the service.

People from the private sector usually understood immediately: how you treat a customer reveals what you believe that person is worth. Aid organisations found it harder.

Twenty years separate the young woman entering Kabul's bazaar from the woman responsible for our work there now, and the distance the world has travelled between them is shorter than either of us would like.

This chapter is far from a complaint. It is an argument for distrusting the assumptions that tell us, before the evidence arrives, who will speak, who will lead, who will stand beside whom — and what a person at the bottom of the pyramid deserves.

The people who fit least neatly into a system's categories are often the easiest to overlook. That does not protect them. It leaves them more exposed to being defrauded, patronised and failed, including by institutions established to help them. They are also the people who live with the consequences of decisions made elsewhere. That makes their account central to the evidence: it is where the decision becomes real.

3

WHEN THE BANKS LEFT

Some time after the 9/11 attacks, when Britain was assembling a new agency to fight serious and organised crime, I sat across a table from part of the team building what became the Serious Organised Crime Agency (SOCA) — today's National Crime Agency (NCA) — and was asked the question I have been answering ever since:

What percentage of *hawala* is criminal?

I thought it was a dumb question. I may even have said so; I was younger then, though silence would probably not have helped — my face has always tended to answer first. A trust network does not have a criminal percentage any more than the English language does. The same dealer, in the same shop, in the same hour, might move a wheat trader's settlement, a refugee family's rent, a UN agency's payroll and — in the wrong town, on the wrong day — something you would want stopped. The useful unit was not *hawala* in the abstract. It was the particular dealer, corridor and relationship.

But the question deserved a better answer than my irritation, and it took me years to see that it was the right question asked in the wrong unit. What they wanted was a way to act proportionately — a way not to treat every money dealer as Osama bin Laden's personal banker.

I entered that room expecting law enforcement to be the obstacle and the aid world to be my natural ally. That assumption would also be overturned. Some investigators proved more willing to test evidence than organisations whose missions rested on compassion. Suspicion, at its best, wanted something it could examine. Good intentions did not always produce the same curiosity.

What the famine proved

Years before the 2011 famine, during another period of hunger in South-Central Somalia, I had been present at a food distribution held in a hot, dusty clearing in the middle of the bush, enclosed only by a makeshift fence. There was barely enough for the people whose names appeared on the list, and nothing for those whose names did not. As the light went, people began coming over the fence. Someone was shot. We lost radio contact with base and drove out through territory held by famished child soldiers. Back at base, I found a colleague nursing his hand: a woman had been so desperate that she had almost bitten through his fingertip as he tried to offload the supplies.

So when famine took hold again in 2011, I needed no briefing to imagine the human reality behind the reports.

I also knew what the appeals rarely showed. In the same country, money sent by taxi drivers in Minneapolis and nurses in Whitechapel was reaching families through non-bank financial institutions (or 'remittance companies') at the speed of a phone call, in amounts they could plan around. Nobody climbs a fence for money that comes addressed to them.

Two years later, the regulated corridor that had outrun the aid system was dying — not from criminality, but from compliance. Britain's last clearing bank was closing the accounts.

I led the assessment commissioned by the British government during that window, and the numbers foreshadowed the decade to come. The regulated UK–

Somalia corridor was moving roughly US$380 million a year, while official statistics captured only US$39.7 million. Remarkably it delivered money at an average cost of 5.14 per cent — already within the G8 target and far below the African average. Closing the accounts did not stop the flows. It made them harder to see.

The assessment proposed what seemed to me the obvious architecture: keep the corridor alive and make it auditable. Subject the operators to independent oversight, screen transactions properly and verify payment at the far end. Build a safer corridor, not an absent one. The logic was the bazaar's, formalised: examine the people, relationships and points of vulnerability you can actually test, rather than pretending the need for money will disappear with a bank account.

Records and reputations

I had begun by trying to help a major player in the Somali remittance industry. I had spent enough time in its offices to know the business was real. Every day a cross-section of the community came through the doors: old men, mothers, students, shopkeepers and taxi drivers carrying small bundles of cash that represented school fees, medicine, rent and gifts to family members they might not see for years. It was plainly a lifeline.

That was why something in the records troubled me. Working through the UK customer data, I noticed that in town after town an improbable number of customers appeared to share exactly the same date of birth. It did not fit the diversity I had seen with my own eyes. Whatever shortcut had produced the pattern, the records no longer reflected the people they were meant to describe.

I raised it privately, as I would have with any client. My advice was simple: fix this before someone else finds it. You cannot ask banks and regulators to trust your systems while your own records contain weaknesses that invite unnecessary questions. It was meant in the spirit of loyalty rather than attack — the conversation you hope happens before the regulator arrives, not after.

The finding was never substantively rebutted. Instead, the response shifted towards managing the person who had raised it. A public-relations firm was hired. Invitations thinned. I found myself excluded from meetings in which my fieldwork continued to be cited. The response reached its most instructive moment on a stage in London, when a Member of Parliament sharing the panel declined, in front of the room, to shake my hand.

Perhaps the people involved believed they were protecting a vital service from a damaging story. But protecting the story did not improve the data, and weak data was precisely what gave banks permission to withdraw. It was another assumption overturned: loyalty did not always mean defending an institution from criticism. Sometimes it meant insisting that its records match the people before the gap became impossible to defend.

The same lesson, at war speed

By 2015, the question had moved to Syria, where the

banking system in opposition-held areas was in a state of complete collapse. The only way to move money to millions of people was through the money men who had stepped into the void — merchants running mobile-phone shops in Kilis, Gaziantep and Reyhanli, connected to counterparts in Aleppo, Sarmada and A'zaz.

We interviewed dealers across four corridors (Iraq, Lebanon, Jordan and Turkey) and the besieged areas of Syria. For near-identical transfers, commissions ranged from 0.03 per cent to 25 per cent — a spread that had less to do with risk than with the fact that no two charities compared notes. Deliveries, meanwhile, were close to flawless. One Islamic charity operating in rural Idleb and Aleppo told us that in nineteen months it had not lost a single dollar to theft or extortion. The transfer system worked. What did not work was everything around it.

The due-diligence standard of the hour was captured in one line of our fieldwork: almost everyone interviewed believed that obtaining three quotes for sourcing and delivering the cash constituted due diligence. Where screening happened at all, charities ran a name — often no more than a rough Latin-script rendering, without the original Arabic spelling — and perhaps an unverified date of birth against a sanctions list, then called it counter-terrorism.

Aid liquidity and war-economy liquidity pooled in the same accounts — the aid wanted in, traders wanted their profits out, and both used the same plumbing. In areas controlled by Da'esh — ISIS, as it was generally known in English — everyone's fuel was Da'esh fuel. In Turkey, the transfer channel on which every agency depended was, strictly speaking, illegal. None of this made the payments unnecessary. It made shallow certainty especially dangerous.

It was in that context that I learned of a meeting I was never supposed to know had taken place. A delegation of law enforcement officers had met officials in the aid department to warn them against engaging a particular merchant, citing serious concerns involving money laundering, people smuggling and terrorism. Yet that same department and the organisation implementing its programme were pressing us to use the merchant on a contract. The department's instruction to the officers, as it was relayed to me, was simple: do not tell Dr Thompson about this meeting.

The officers were bewildered. They saw me as the 'aid person' — the humanitarian counterpart they knew best, working to protect both the programme and national security. They could not understand why the aid department wanted the warning kept from me.

We refused the merchant anyway, on the strength of our own due diligence. We lost the contract. The implementing organisation nevertheless routed the payment through this merchant's network, and it produced the equivalent of £1 million locally from the same liquidity pool whose provenance had prompted the warning.

The problem was not an absence of information. It was an absence of ownership. No single person was accountable for carrying the warning through to the decision, recording why it had been accepted or rejected, or answering for what happened next. The risk could be passed from desk to desk until it had nowhere clear to land. The merchant received the money; the taxpayer carried the exposure; and responsibility evaporated between the institutions involved.

The same silhouette from the outside

The police had warned us away from one merchant. Another would become indispensable to our work in Syria. From the outside, the two men looked much the same: cash, influence, discretion and cross-border networks.

Every money market I have worked in has had one: a person whose name is spoken differently. Not louder — quieter. A market has a broad middle, and then one or two figures near the top through whom, somehow, everything eventually passes. Outsiders often assume the whisper means criminality. Sometimes it does. Sometimes it means something more complicated: competence, reach and a reputation tested under pressure.

Syria's was a merchant whose name I still say with a certain gravitas. The textbook on men in his position is not kind, and the textbook is often right. I know what it can cost to get too close to the centre of a network, and I do not recommend making a habit of it. I arrived at a different judgement about this man slowly, through some very hard periods in the war. Hard times are the most expensive due diligence there is. Without him, the first delivery of Amanacard cards would never have crossed the border into Syria.

My father had designed a small number of commemorative coins for people whose work deserved recognition: the Amana symbol on one face, the words 'unsung hero' on the other. They had almost no monetary value. Their purpose was to tell someone, properly, that what they had done had been seen. I carried two with the first shipment of cards.

The cards left London in beautifully ordered boxes, sequenced by serial number. The merchant looked at my system and threw the lot into black bin bags — the sort of luggage that could cross the border without attracting attention. I later learned that he had searched the bags himself, found the two coins and arranged a separate armoured vehicle to carry them across.

The episode was funny, and then less funny the longer I thought about it. The operational cards had travelled disguised as rubbish because their value lay in what they would do. The coins travelled under armed guard because their value lay in what they said. The metal was incidental. What needed protection was recognition: a small, physical acknowledgement that someone had carried risk faithfully and had been seen. I had spent years studying money and still needed a merchant to remind me that metal has weight, a card has authority only when a system will honour it, and trust is what makes either one move.

Place him beside the merchant the police had warned us about and the SOCA question becomes concrete. One presented a risk we would not accept. The other kept his promises when the formal systems around him had ceased to function.

No sanctions screen could settle that distinction from a distance. Nor was intuition enough. Judgement had to be made close up, but it also had to be recorded, challenged and made answerable to evidence. The criminal percentage of either man was not a number. The question was what this person had done, through which relationships, under whose observation, and what happened when something went wrong.

The hub that never was

By then we had seen the same failure twice. Banks closed corridors because they could not see through them, while aid organisations repeatedly bought shallow fragments of diligence instead of building shared knowledge. The remedy we proposed in 2015 is still on the public record: donors should pool funds to create a central know-your-customer and know-your-supplier processor for all their partners.

The hub would be staffed by people with real country and financial expertise. It would check remitters and recipients against sanctions lists and the deeper sources available to law enforcement; maintain a shared risk matrix; and preserve what had been learned about each dealer, network and border. Instead of every NGO buying the same thin review, grant after grant and donor after donor, the difficult thinking would be done once, to a high standard, and kept current. Enhanced due diligence in a war, we argued, was a proper use of charitable funds — a line item, not an embarrassment.

The institution that understood the proposal most readily was, to my surprise, Chatham House. When its Middle East programme convened a working group on supporting Syrians without supporting the war economy, the people around the table had read the assessment and immediately recognised the missing discipline: the financial supply chain. A think tank and the police became two of the proposal's strongest allies.

The wider aid system remained fragmented. Knowledge disappeared when a grant ended, a consultant left or a report was filed. The arrangement looked cautious because everyone held a piece. In practice, nobody held the whole picture.

Had the hub been built, donors could have spent less while learning more. Every implementing partner in a difficult corridor could have drawn on one continuously maintained body of knowledge: this dealer, this network, this border, this warning, this successful delivery. Instead, a decade later, institutions still repurchase fragments of understanding and call the repetition assurance.

Collins Street

Twenty-five years after the coin was flipped, the flip found me at home. The day before I flew to Papua New Guinea in late 2025, I decided to stop treating China as the great geopolitical unknown — discussed endlessly as a strategic threat, encountered mostly at a distance — and walked into a Chinese state bank's branch in Melbourne's Chinatown to ask whether it would bank my new Australian remittance business.

I was at my least glamorous. I had come straight from a walk around the Royal Botanic Gardens, in tired gym gear, working down the list of jobs before packing — at which point I had remembered that, among everything else, we still did not have a bank. I am not, as a rule, a woman who meets her account manager without makeup and a suit. I was also carrying an enormous brown paper bag marked Sportsgirl, laden with presents for women I had met in Port Moresby — sunglasses and fun costume jewellery, ordered by request.

The security guard was unsure what to do with me, called ahead and sent me to the head office, where I was eventually escorted — gym gear, Sportsgirl bag and all — to the top floor of a very handsome building on Collins Street, where the shopping bags are usually Gucci or Chanel from the boutiques downstairs.

The man who met me was polite and direct. I was already slightly surprised that the meeting was happening at all. I had walked in unannounced at street level, looking nothing like the sort of customer a Chinese state bank might reasonably expect to find interesting, and somehow been admitted to the top floor.

Once there, however, the conversation sounded remarkably familiar. From one brief telephone message he had already formed an accurate picture — charities, remittance companies, trade in difficult markets — and the bank had no appetite for that kind of risk. It was conventional banking logic, delivered unusually plainly.

I mentioned that I had seen his bank's presence in Papua New Guinea, where I needed to trade. He looked genuinely shocked. They were all terrified of that country, he said. Nobody left the hotel. I happened to know the hotel.

Then he offered another familiar piece of banking advice. The conversation might be different, he said, if I came with a trusted recommendation from someone in mainland China.

The recommendation would not make the regulatory questions disappear. It might make the bank willing to look at them again.

That was what struck me. In the space of one meeting I had encountered the same mechanism I had seen in the bazaar and in banks across the world: rules established the boundary, but relationships often determined whether somebody was willing to open the door, look more closely and make a judgement.

In Kabul, an introduction got you into the room; what happened afterwards determined whether you were trusted. On Collins Street, the architecture was entirely different, but the human logic was recognisable.

Who knows you? Who will vouch for you? Is there enough trust for the conversation to begin?

In the lift down, a group from a large Western bank stepped in, polished and entirely at home in the building. Their institution would have used different language. One banker had once advised me, however, not to tell his bank too much about the nature or destination of my payments.

Too little information is suspicious. Too much reality is inconvenient.

The proposed hub was an attempt to live in that uncomfortable space: enough information to make a payment accountable, without pretending the world at the other end was tidy.

Build it anyway

Nobody built the hub. In the end, I stopped proposing the architecture and built a smaller version of it privately, in the hardest corridors first. Amanacard grew from the same findings: verify the person receiving the payment; use named local agents; record the transaction so that a bank, donor or investigator can follow it; preserve the human judgement rather than replacing it; and make knowledge reusable instead of letting it vanish at the end of each payment round.

The design was not invented in a product workshop. It was extracted, finding by finding, from a famine corridor the banks abandoned and a war the banks never entered.

And the SOCA question — the dumb question, the right question — finally has my grown-up answer. You cannot know what percentage of a trust network is criminal, and you do not need to. You need honest activity to be visible enough to defend, and suspicious activity visible enough to test.

That does not mean exposing everyone to everyone. An Afghan girl may need to disappear from the Taliban while the payment supporting her remains visible to the bank overseas. Privacy can protect the person; accountability should still follow the money.

The real question is not whether something remains in the dark, but from whom. Build a system in which legitimate activity can be examined by those entitled to examine it, without making vulnerable people visible to those who would harm them, and the percentage question begins to lose its power.

The banks left. The money did not. What they left behind was not an empty corridor, but an unobserved one. The work began there.

4

THE UPSIDE-DOWN PLANET

To understand where the later work began, I need to step back from the Middle East to an earlier natural disaster, before the card had a name or I knew what it needed to do.

In 2013, the British government funded an experiment: the UN's first dedicated Cash Coordinator. I was deployed into the wreckage of Super Typhoon Yolanda in the Philippines.

Tacloban felt like Mars. Cars jutted from rooftops, as though they had been lifted mid-journey and dropped nose-first from the sky. The dead lay along the roadsides. The streets were scrambled beyond comprehension, piled with timber, metal, furniture and household things like a town-sized junkyard in a power cut. More than fourteen million people had been affected and more than four million displaced, but numbers could not describe the physical inversion of the place. Things built to remain on the ground were above it. Things meant to protect people had collapsed over them.

And through it all, Filipinos were quietly clearing, repairing and starting again. They did not wait to be told that they were resilient. They had work to do.

I arrived with none of the tools I would later build and a job title that had been invented more quickly than the reporting line around it. Cash had been used in emergencies before, including in the Philippines. What was new was the attempt to coordinate it across an entire international response: government, UN agencies, charities, banks, remittance companies, mobile networks and the communities receiving the money.

Today, cash working groups are standard enough to sound

inevitable in an aid response. They were not inevitable then.

A job without a home

Cash crossed every institutional boundary. It could be food, shelter, livelihood support, emergency employment, social protection or the means by which a displaced family returned home. That made it everyone's business and, administratively, nobody's.

I was inside the UN this time, rather than advising it from outside. It was one of the hardest jobs I have ever done. The storm had shown no respect for organisational charts; the international system clung to them. There were meetings about whom I reported to, meetings with people who believed I reported to them, and whole parts of the response I was expected to coordinate without the authority to bring them into the same room.

I was asked to create common guidance across organisations whose programmes, mandates and budgets all approached cash from different directions. Cash also unsettled a deeper assumption. Much of the aid system had been built to procure, ship, store and distribute goods or deliver services. Cash handed some of that choice back to affected people and local markets. That made it practical, but also disruptive.

The resistance was not because the people involved did not care. Most were exhausted, working at speed and carrying impossible loads. The difficulty was structural. Cash exposed the joins between UN agencies, and institutions are often most defensive when a 'new tool' seems to encroach on their territory.

I reached one of my lowest points there. The work was relentless, the politics dispiriting and the level of effort unsustainable if evenings and weekends were meant to exist. For a while, it felt as though the people who most needed coordinating were those least willing to be coordinated.

The reprieve came from the edges of the system.

The small mapping team within UNOCHA, the UN's crisis-coordination office, was one of the least celebrated and most useful groups in the response. They did not need to win an argument about ownership before making something visible. Give them good information and they could show, on one screen or in one of their beautiful print-outs, where the storm had hit, who had already been poor before it — and therefore had the least room to absorb another blow — who had now been pushed below the poverty line, where people had been displaced, what assistance was planned and which roads, phone networks and payment points were beginning to work again.

The team's weakness was not competence. It was timing. Too often the maps and briefing notes arrived late, not because the work was slow, but because the bureaucracy around the team pulled its attention upwards: reporting into the system rather than looking outwards at what affected communities needed.

The Philippine government meetings were another corrective. I had the rare pleasure of sitting in coordination rooms in which almost every government official was a woman. They were direct, demanding and protective of their people. They were not there to admire the machinery of the international response. They wanted to know where the gaps were, why aid organisations were duplicating one another and whether our plans reinforced the government systems that already existed.

I had to work hard to earn their trust. That was as it should have been. They were accountable to the people around them long after the international staff had rotated out.

Cash is not one thing

The argument for cash begins with a simple act of respect. If food, roofing material, medicine and school supplies are available locally, the person who has lost a home will know better than a distant international organisation which of them is needed first.

Goods can save lives when supply chains have genuinely broken or when time is measured in hours. Cash cannot conjure rice into a sealed-off town, make a dangerous journey safe or hold down prices in a market being deliberately exploited. But markets are rarely as absent as outsiders think.

In crisis situations, entrepreneurship survives, and can even thrive: merchants find routes, families know who can source what, and small traders move as soon as there is demand worth serving. Cash gives them that chance. It restores purchasing power to the people affected, and that purchasing power pulls goods back through the local system. One family may need corrugated iron. Another may need antibiotics. A third may need a bus fare to reach relatives, replace the tools of a trade or repay the neighbour who carried them through the first week.

The technical language made cash sound more complicated than it was. One decision concerned purpose: unrestricted cash, cash linked to an agreed activity, or vouchers limited to a shop, category or amount. Another concerned delivery: banknotes in hand, a remittance counter, a mobile wallet or a prepaid card.

Cash and cards were not opposites. In humanitarian work, cash meant the value a person could use; a code, phone or card could establish the entitlement, release that value and, if the system was well designed, record what happened.

It helps to see the whole journey of a single grant, because each stage looks reassuringly modern until the last one. A donor transfers funds to an aid organisation; the organisation transfers them to the headquarters of a bank, remittance company or mobile operator; the provider allocates value to a list of names and pushes it out to branches, merchants or wallets. Every step so far is electronic, auditable and quick. Then the journey leaves the diagram: a human being has to stand at a counter in a damaged town and hand over physical money to another human being, in the right amount, to the right person, with a record that survives. The whole architecture is only as strong as that final metre.

Each mechanism moved the difficulty elsewhere. Direct distribution required discreet transport, security, counting and reconciliation. Remittance counters were familiar but imposed repeated journeys and fees. Mobile wallets depended on phones, signal, literacy and a merchant with enough cash. Prepaid cards depended on powered, connected machines and were often discarded after a single programme.

Each mechanism solved the transaction in front of us. It built almost nothing beyond it. The records belonged mostly to the charities, banks or service providers. They showed that a payment had been made, which mattered, but they did not usually become a credential the person could carry into the next month, the next grant, the next employer, the next clinic or the next displacement.

A payment could be digital at every stage in Manila and still fail in Tacloban because nobody at the far end had pesos to hand over. Liquidity is a bloodless word for a very human fact: someone has to have the cash when the person entitled to it arrives.

The pawnshops

I had always had a faint prejudice against pawnshops. They were familiar enough — part of the streetscape, almost a neighbourhood fixture — but in my mind they belonged to the dirty side of finance. Their windows suggested desperation: jewellery surrendered through a grille, need made visible and priced. I thought of them as ordinary and faintly shameful at the same time.

That was my association, not the evidence.

In the Philippines, pawnshops were woven into everyday financial life. People collected remittances there, paid bills and received money from family members working elsewhere. Some were as familiar as the *sari-sari* store, the local equivalent of the small family-run shop. They were not an exotic substitute invented for a disaster. They were part of the infrastructure people already knew.

One pawnshop network we assessed had 1,700 branches across the country. A charity could upload a list of names. The recipient arrived with a transaction code and an accepted form of identification; where formal identification was unavailable, confirmation from the local *barangay* — the smallest unit of government, roughly equivalent to a village — or an NGO could be accepted. The branch logged the time, place and amount in real time. The recipient signed for the cash. This was not money disappearing into a back room; it was an auditable payment through a familiar counter. But accountability still sat around the transaction rather than inside it: paperwork, sampled checks and fallible humans reviewing other fallible humans after the event.

In Tacloban, the network was rebuilding branches from nothing. The banks could not provide enough cash, so the network helicoptered banknotes into the city itself.

That image rearranged the categories in my head. The pawnshop was flying in the liquidity on which the formal response depended — not the banks.

I had spent years objecting when Western institutions treated *hawala* as suspicious because its shops and relationships did not look like banking. Yet I had made the same error myself, confusing the appearance of pawnshops with evidence about how they worked.

The useful question was not whether a financial counter looked respectable to me. It was whether it could deliver, protect the person collecting, maintain enough cash, keep an honest record and answer when something went wrong. On those tests, some of the pawnshops were performing remarkably well.

They were also a reminder that the last mile is rarely served by the institution with the grandest headquarters. It is served by the outlet that is still open.

The map

Cash programming did not begin when the banknotes arrived. The chain we designed began before the money moved.

A charity began with the government's reconstruction priorities, then worked with the mayor to identify gaps in coverage. At *barangay* level, the local captain and an independent community council — representing men and women of all ages — identified those most in need.

The first list was often handwritten. We pushed to match it against government data and keep it current. Once the list was confirmed, the woman in the household was normally named as the person receiving the grant. The community was told who qualified and why, and the charity returned after payment to find out what had actually happened. At the time, I did not realise how consequential that choice could be: to make the woman visible in the financial record, rather than defaulting to the man as head of household and therefore the person through whom the household's money was held.

The process was imperfect. Local lists can exclude people. Government data can be old. A woman named as the grantee may not be able to travel safely. A payment point marked as open may have no cash by lunchtime. But the chain made each judgement visible enough to question. It connected the handwritten list to the national picture without pretending that one was more real than the other.

With the mapping team, I began layering pre-typhoon poverty data with the path of the storm, displacement reports, planned assistance and the recovery of financial and communications infrastructure. The map could show not only where need was greatest, but where a proposed mechanism might work.

This changed the question. Instead of asking whether mobile money was modern, we could ask whether the network was working in the municipality where people had been displaced. Instead of admiring the reach of a card scheme, we could see whether any ATM near the community had electricity and cash. Instead of letting five charities announce similar programmes in the same place, we could show the gaps nearby.

Maps turn institutional claims into geography. A programme described as nationwide becomes a scatter of dots. Coverage becomes the distance a woman must actually travel.

The private sector responded to that kind of evidence with an energy I found exhilarating. The humanitarian working groups contained good people, but also some who had seen so many emergencies that every proposed change reminded them of an earlier failure. The Philippine business community had a different rhythm. Telecoms companies, remittance networks, consumer businesses and banks were already restoring services because their customers needed them and their businesses depended on it.

I made a new friend, Jill, who worked for Smart Communications. She drove me around to a soundtrack of upbeat Christian worship music and introduced me to a business community that wanted to solve practical problems. We used the maps with Smart to identify places where restoring coverage would reopen not only communication but payment. A live phone mast could reconnect families, notify a recipient and allow a local merchant to authenticate a cash-out. Communications infrastructure and financial infrastructure were becoming the same map.

For a few weeks, at the peak of the disaster, government, business and the aid system briefly rowed in the same direction.

The island

Jill also gave me something the coordination rooms could not: a view of the social world that remained intact around the disaster.

She took me to her uncle's farm on Cebu. Local families came in for a community distribution of goods. Nobody performed humanitarianism. It looked like what it was: people with something to share helping neighbours who needed it.

The mangoes were the most succulent I have ever tasted. One mouthful transported me to Melbourne as a child when Ros Feeney, the mother of my friend next door, showed me for the first time how to cut a mango cheek into squares and push the skin inside out. It felt marvellous then: a small domestic magic trick. Memory works strangely in emergencies. A taste can return you, without warning, to the safest place you have ever known.

That night they projected *Gravity* outdoors. On the screen, Sandra Bullock and George Clooney floated in black space; around us, fireflies sparked and vanished in the dark. For a few hours, we seemed suspended too — in a small, alternate universe of peace, with the damaged world waiting beyond the circle of light.

The international system often describes people after a disaster by what they lack: shelter, food, income, protection. All of that matters. But the farm reminded me how much survives the inventory. Family, faith, humour, generosity, music, memory and the instinct to help before a programme has been approved.

Cash works best when it respects that surviving world. It does not require a family to reorganise its priorities around the contents of somebody else's relief package. It can travel through businesses people already use and support the market trying to stand up around them. It treats the recipient as a person making a decision, not merely the destination of a commodity.

What I had not yet understood was how narrowly I was still thinking about cash. In the Philippines, I saw it mainly as support for families: money for food, shelter, medicine, transport and the first steps back to ordinary life. Only later, looking back across the war zones I had worked in and then returning to them, did I see the larger pattern. When banks retreat or collapse, cash is not one humanitarian tool among many. It is the operating system of almost everything.

Hospitals pay doctors and nurses in cash. Schools pay teachers in cash. Debris is cleared by workers paid in cash. Ambulances are repaired, buildings rented, meals catered and supplies bought from vendors who also expect cash. The aid system placed household assistance in a box called cash programming, while salaries, procurement and operations sat in separate boxes. Yet on the ground they all depended on the same thing: getting value safely to the right person or business, and being able to show that it arrived.

That was the larger opportunity: a safer corridor that could serve all of them. As aid budgets contract, the system can no longer afford to rebuild a different payment chain for every institutional category.

The missing record

The Philippines gave me the distinction I would carry into everything that followed.

Cash is not secure in every sense. It can be stolen, extorted or lost. Moving it in bulk creates obvious danger. In conflict zones, it was normally carried discreetly in ordinary cars, not armoured vans. When the local currency had collapsed, however, paying salaries or suppliers could require bags of low-value notes, filling a vehicle and making concealment almost impossible. US dollars were often preferred because they retained their value and reduced the physical volume, although using them brought US Treasury compliance directly into the payment chain. None of this altered the underlying reality: when institutions fail, electricity disappears and networks go down, cash remains the most resilient form of value a person can hold. It requires no battery, signal, merchant terminal or institution's continuing permission.

Identity may be the last credential a displaced person still possesses. Cash is the value that identity can unlock, restoring some room for choice and dignity.

The design problem was not how to eliminate cash. It was how to preserve its freedom and resilience while creating an accountable record of delivery.

I did not move directly from that insight to building Amanacard. I first tried the obvious route: persuade companies that already moved remittances to give the idea a home.

In 2014, I took what I had learned in the Philippines to a large remittance firm. The proposal was simple: use existing payout relationships in disaster-affected countries as a shared wholesale channel for humanitarian organisations, instead of every charity rebuilding the same network in the middle of an emergency. The discussions went far enough for the firm to offer me a formal role reporting to its chief executive. But fragile states were hard to fit into the growth story its investors needed to hear, and the humanitarian product never reached prototype.

Not long after, I met the same limit with another remittance firm. It had an extraordinary merchant network, but making fragile states central to strategy was another matter. There was an irony in this. Many remittance companies began with migrant counters, family shops and trusted neighbourhood agents; as headquarters grew, strategy moved upwards too. A single local bank looked cleaner than a pawnshop, merchant counter or family-run shop. But the useful question was still the old one: who could actually release value to the right person, in the right place, with a record that would stand up afterwards?

I am glad I tried. The exercise forced me to separate the parts of the problem. Existing remittance rails could move value quickly across a border. What the aid system still lacked was the verified last mile: a reliable connection between identity, entitlement, the named person releasing the cash and the record showing that the right person had received it.

Existing firms could move money into these markets through intermediaries while keeping the people at the far end at arm's length. None was ready to make the recipient the starting point.

So we went alone.

A remittance business begins with the sender and reaches towards a recipient. Amanacard would have to begin at the other end and build back from the person receiving. That distinction became part of the design.

The card we later built does not follow individual banknotes through a market. It links a person's verified identity to an entitlement. At the point of payment, a merchant verifies the recipient, scans their card, releases the cash and creates the record: who received what, where, when and through whom.

An Amanacard field officer is there too — not as a guard or cashier, but to solve problems, maintain trust and verify that the human activity matches the electronic record.

For the recipient, it is still cash. It can buy medicine from one shop and roofing nails from another. But the transaction can also leave something useful behind: a record that this person was verified, paid and able to transact. For someone otherwise invisible to formal systems, that record may later matter: for another payment, a referral, a grant, a job, a clinic or a bank account. For the organisation funding it, the handover no longer disappears at the moment the bank transfer ends.

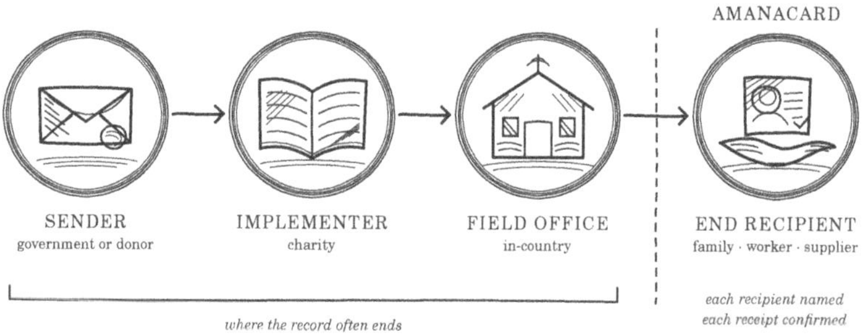

Simplified from the SecureAid money-flow schematic prepared for the German Government, 2018.

At the time, cards were usually presented as an alternative to cash — a cleaner, more modern future in which banknotes would gradually vanish. I came to the opposite conclusion. The innovation was not to make cash disappear, but to make its delivery visible without controlling what happened to every note afterwards.

That principle followed us from the Philippines and Somalia into Syria and Afghanistan. In Syria, the first system my team built was not for household grants but for salaries and supplier payments for hospitals. A card could be the gateway, identity the anchor and cash the usable value. The mechanism changed with the conditions. The obligation did not: the right person or business should receive what they were owed, and somebody should be able to prove that it happened.

But mechanisms were not what carried me through that deployment. People did.

Christmas Eve

By Christmas, I wanted one night off.

I found a tiny hotel, improbably intact. Only the family suite remained, and I could not afford it. I explained what I was doing in the Philippines. The person at reception said the manager would call me back.

The call came. The manager wanted to give me the room.

I arrived on Christmas Eve covered in dust, checked in and went to my room. Within the hour I had showered, washed my hair and made myself look vaguely human again. When I came down to the restaurant, the staff politely turned me away. It was for hotel guests only.

They could not believe that the clean woman in front of them was the creature who had checked in sixty minutes earlier.

Once the confusion was resolved, they looked after me like angels. When I returned to the room, there was a note beside a delicate flower and a candle, thanking me for helping at such a time.

I sat on the bed and cried — not from exhaustion this time. On any other night it would have been kindness, but on Christmas Eve, after weeks of moving through the Visayas, sleeping in tents and whatever basic accommodation was available, the simple fact of being given a proper room felt like grace.

In the morning, I was ready to go back into the thick of it. The institutions were upside down. The people never were.

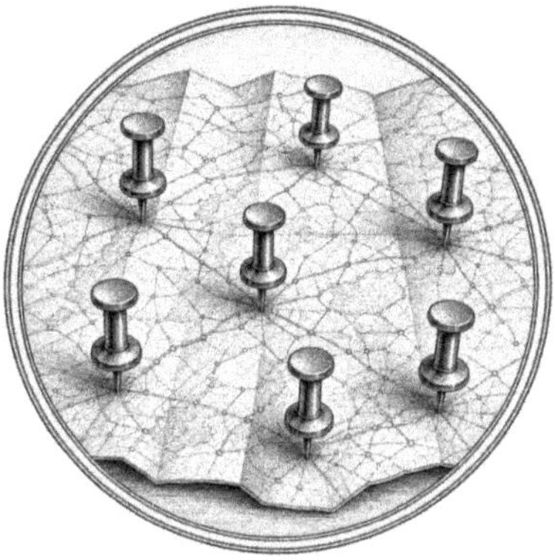

5

THE SEVEN

Days before the biggest and most sensitive operation we had ever attempted, I fired my entire field verification team.

In 2018, we were about to make sure thousands of hospital workers across dozens of hospitals in opposition-held Syria received months of back pay — under Russian air attack, with communications at constant risk of interception by a regime that wanted many of those workers dead. A senior humanitarian worker and a relationship manager at a major UK bank had put their reputations behind the proposition that we could deliver something nobody else had managed.

I had no contingency plan, no replacement team waiting and no clear route out of the problem I had just created. The final trigger came at the eleventh hour, when the man coordinating the expansion refused to let me speak directly to the new zonal supervisors he had recruited. Thousands of people were about to depend on judgements made by people I had not been permitted to assess for myself. It was not an ordinary disagreement about control. The operation had become hostage to one man's refusal: proceed on his terms or not at all. I remember thinking, with more exasperation than diplomacy, that we did not negotiate with terrorists. In work where a weak judgement could expose thousands of people, keeping the wrong team felt more dangerous than having no team at all.

To understand how I had learned to notice that — and how the answer walked into my life the following week — you have to go back to a very young interpreter in Afghanistan.

Moslem

I never trained my first verification agent. We trained each other.

Precocious and quick, Moslem Shah became my voice into Dari and Pashto as we travelled Afghanistan interviewing the same kinds of men in city after city. But the interviews were only half the work. The conversations afterwards were often more valuable: what had been said, what had pointedly not been said, and what the seating, silences and interruptions had meant.

Those debriefs built our picture of the money men slowly, layer by layer. They taught me how to work across language without mistaking translation for understanding — and how to listen as carefully to what was not translated as to what was.

Years later, that way of listening became the skill on which everything else depended.

Saad

At a dinner in London with friends I had first met in Afghanistan, I described the problem in Syria. I had already made several failed attempts to build a verification team and had learned that capable people were not difficult to find. People with the right instincts, values and willingness to be accountable were much rarer.

I was looking for someone to lead a small team and tell me what they saw rather than what they thought I wanted to hear, even while their homes, livelihoods and communities disappeared around them.

One of the friends at the table was Tom, who had been *The Economist*'s correspondent in Afghanistan in 2005 and had since reported on the Syria conflict. He said he knew one man whose integrity and judgement he trusted without reservation. He had seen him under extreme pressure.

His name was Saad.

Tom had been with Saad when a regime sniper shot Saad's father in the thigh. They struggled to get him to a field hospital while he bled heavily; after several days, a shortage of beds meant they had to take him home to recover. He survived the shooting. Months later, Saad's father was killed when an air strike destroyed the family home.

I asked for his number and met him over Skype the next day. His English was broken. My Arabic was worse than broken. It did not matter. I knew almost immediately that I could build something with him.

I gave him one task.

Come back tomorrow with seven CVs.

Not seven people like him. Seven completely different people — teachers, bakers, horticulturalists, engineers; I did not mind. Everyone notices different things, and I wanted eyes that noticed differently. If he already knew and trusted some of them, so much the better — that was how the bazaar hired too — but he was not to bring me his extended family.

The next day we interviewed them together, in broken English between us and Arabic among themselves: extrovert and painfully shy, instinctive and methodical, one of them surely the tallest man in Syria at very nearly seven feet. They could hardly have been less alike, yet somehow they fitted together perfectly.

Saad recruited trust the way the bazaar always had. The seven then proved him right under aerial bombardment.

Here was my own thesis, performing itself on my own payroll.

Trust made the group possible. Difference changed what the group could see.

Keeping the core field team at seven was deliberate. Partly it was budget: we wanted to prove that enormous things could be done with a tiny team, leaving as much of the aid dollar as possible for the people it was intended to reach. But small teams also learn together. You can watch their judgement change as they compare what each person has noticed.

The seven were not the whole operation. Alongside them was a small, revolving core team in London that became its backbone: refining the design against what Syria taught us, solving the problems that arrived each day and turning an improvised field method into something that could actually run.

Together, we were replicating something I had studied in the bazaar without understanding that I would one day need it: a small trusted network in which reputation travelled faster than paperwork, but decisions could still be recorded, challenged and examined.

The shape of that network first appeared to me as colours on a map.

The colours

We began with almost nothing: a list of hospital names and

a rough idea of where they might be.

The seven's first assignment was not to verify anyone. It was to find the hospitals, map them and organise the work. We divided the country into coloured operational zones, choosing each colour to honour what those places had meant before the war rather than allowing the war to define them.

When we later expanded into the north-east, we chose a warm wheat gold because it was Syria's breadbasket. To this day, I think of each member of the original team by the colour of the region they first managed.

The team told me the green zone was the most beautiful part of Syria. Saad occasionally took them and their families there for picnics by the river, giving everyone a circuit-breaker from the intensity of the work.

Once, he took them to a swimming pool in the coral zone. My father followed the Arabic WhatsApp groups through translation software which, despite all its cleverness, informed him that the team had taken the day off to "drown their children".

"What on earth is Saad doing with the team today?" he asked.

After the regime fell, we finally brought the enlarged team together from across Syria in January 2025. By then, the core seven had become eleven. One of the first places I wanted to see was Darkoush and the river I had heard about for so many years. We stood there in stunned silence with one of our favourite hospital administrators, looking at a place that until then had existed for me through maps, hospital lists, schools, markets and years of detailed logistics planning.

Travelling farther south, I noticed how many hotels, cafés and businesses carried the name Jasmine. Only then did I understand why the team had chosen it for our work in the besieged and regime-held areas — the operations kept most secret because discovery could cost lives. Jasmine was not simply a pleasant name. It was how Syria remembered itself.

The colours on my planning board had always been real places to them. They had not been choosing colours. They had been choosing home.

Go and build trust

The day after the first map was made, I gave them the real assignment: go and build trust.

They did not have long. Thousands of people needed to hand complete strangers one of their most valuable remaining possessions — their identity — in exchange for a promise that we would protect it, and a small card that would become their credential for every salary and supplier payment to come.

That card would unlock months of unpaid wages.

At the same time, the team had to win the confidence of the merchants who could supply liquidity around each hospital. They approached both groups in the same way: slowly, respectfully and by listening far more than they talked.

Amana carries in Arabic the sense of trust and custodianship: the obligation to safeguard what has been entrusted to you. The seven were being entrusted with identities, and they worked as though the name were an instruction.

That was where I really learned what verification means.

It is not the collection of as much information as possible. It is building enough understanding to establish confidence while interfering as little as possible in lives that are already hard beyond telling. That was know-your-customer ('KYC') in its most literal sense — not the accumulation of documents to satisfy a distant checklist, but the work of actually knowing enough.

Compliance became intensely local because the risks were local. In some places the dominant threat was regime surveillance; elsewhere it was al-Qaeda or Da'esh. Around them moved the ordinary predators of any war: smugglers of people, weapons, drugs and goods.

A merchant's relationship to the regime — whether he had benefited from it, resisted it, lost family to it or shifted allegiance as control changed — formed part of the context, never a rule. The point was to understand incentives and loyalties rather than allow a generic checklist to pretend it had done so.

Identity also had to be rebuilt from first principles.

The documents assumed by compliance textbooks — current, verifiable and issued by trusted authorities — often did not exist. They had expired, been abandoned in flight or been issued by institutions that nobody trusted any more.

So we stopped asking:

Do you have the correct document?

And began asking:

What evidence still carries trust here?

Employment records. Hospital payrolls. Merchant knowledge. Repeated interactions. Community reputation. Consistency over time.

Identity became a layered judgement rather than a single document. That quiet inversion became one of the foundations of everything Amanacard later built.

The price of seven

A team as small as seven creates extraordinary trust, and it terrified me. Every individual mattered enormously.

One day, six of the seven became trapped simultaneously in hospital basements under heavy bombardment. They had ignored the operational plan and all gone out at once to facilitate and verify the payroll payments. For several hours I could not reach any of them, and I believed I had lost almost the entire team.

I have rarely prayed as hard as I did that day. They all survived.

Another time, the man we relied on most in north-east Syria was kidnapped.

The call came while I was meeting Amanacard's then head of special projects. He knew more about captivity than anyone should: ISIS had held him for more than a year alongside British and American hostages who became his friends in the darkness, many of whom did not come home.

I count his return among the great strokes of fortune in my life. Without him, there would have been considerably less humour in the aid dramas, and a good part of what we built would not exist. I had seen extraordinary resilience in people living through war, where simply getting through the day could demand it. His was different. He had returned to an otherwise ordinary, privileged life after witnessing some of the worst things human beings can do to one another, and somehow remained relentlessly positive.

That was why I understood the weight of the moment when, for the first time since we had met, he quietly asked to leave the room.

It was too close.

The kidnapping in Syria continued. From London, I worked every channel available to me, and the family of the man who had been taken knew we were trying. Others, much closer to the situation and carrying the real burden of securing his release, eventually brought him home.

That early team was second to none. Every one of us carried a private history of trauma, yet the work was met with extraordinary generosity, humour and determination.

When young people interview with us today, I have to remind myself not to measure them against a team formed under conditions I would never wish on anyone. What human beings can endure while remaining optimistic and open-hearted still astonishes me.

They taught me at least as much as I ever taught them.

The team behind the team

Turning the field method into a system was only the beginning. The London team and our software engineers then kept it working under conditions no ordinary product team would have accepted as normal. From 2016, for the next six years, urgent feature requests and bug fixes arrived at almost any hour. Saturday morning acquired its own small dread: the phone ringing, somebody unable to complete a payment, engineers opening laptops while families waited, and everyone helping willingly because there were real people at the far end of the problem. Weekends were interrupted so routinely that only when the interruptions finally stopped did we realise how completely we had organised our lives around them.

The weekends were only the part you could see. Beneath them ran a state of alertness that never switched off. Through the autumn of 2018, I woke every morning not knowing whether one of the hospitals we were supporting had been hit overnight and, in the worst imagining, by what. The deepest dread was chemical attack: once President Obama's red line, by then entangled in diplomacy, denial and arguments over proof. Images arrived through WhatsApp from inside Syria while the official world debated what could be known.

At the 2019 Cannes Film Festival, Dr Hamza — our operations director — and his wife, Waad al-Kateab, carried the reality of Syria's hospitals onto one of the world's most visible stages. Two photographs from Cannes capture both the pride and the cost of the work for me. Waad had shot some five hundred hours of raw footage. Editors Chloë Lambourne and Simon McMahon worked through every hour and, with Waad and co-director Edward Watts, shaped the footage into *For Sama*, a masterpiece about Waad and Hamza's family and the hospital Hamza kept running during the siege of Aleppo. The film would win the festival's prestigious *L'Œil d'Or*.

In the first photograph, Hamza, Waad and Ed stand on the red carpet holding a sign that reads STOP BOMBING HOSPITALS. Amid the glamour, the sign was startlingly out of place, which was precisely the point. The appeal was not abstract to us: we were then helping pay staff and suppliers across dozens of hospitals.

In the other photograph, taken after the award, Hamza sits amid the celebrations, the trophy beside him — and his computer open on his lap as he solves an urgent problem for Saad in Syria. One photograph records what we were trying to do; the other, what doing it required.

Developers came and went over the years, as they do everywhere. What surprised me was how many wanted to stay, and why. Örvar, our senior solutions architect based in Scandinavia, spent the better part of a decade building the platform's deep machinery, and when he finally moved on he wrote a farewell note I have kept: "It's been one hell of a ride and I am really glad to have been a part of it. If only what I have built has made a difference for one human in the Middle East, then I'm more than content." Engineers are not supposed to talk like that about enterprise software. Ours did, because the thing on their screens on a Saturday morning was never an abstraction. It was someone's salary, someone's hospital, someone waiting.

Across those six years, I remember only one occasion when there was genuinely nothing we could do. I was on the Tube in London, receiving sporadic updates from our chief technology officer, Clive, who had found that the fault lay with Cloudflare, part of the internet infrastructure our software relied on. Hundreds of people were unable to receive their aid. Cloudflare was down. So were we. No amount of ingenuity, urgency or goodwill could bring it back.

In Syria, the team improvised. Yaser, who managed the area we called the blue zone, bought ice creams for the children waiting at the venue. Abdullah Othman used the delay to launch into one of his brilliant impromptu lessons in financial literacy — part instruction and, by all accounts, part light entertainment. A few hours later, Cloudflare returned; the system came back with it, and the payments continued.

What the network knew

By 2020, the eleven members of the core team each had a tightly defined role within a wider network of dozens of contractors and partners. One of those contractors, from north-east Syria, is now a core member of the team and is experimenting with AI tools to help us share occasional glimpses, through social media, of work we once had to keep almost entirely hidden.

There is something fitting in that: people who disappeared inside the work are now helping to make it visible.

Only later did we understand what that structure had quietly protected. No single person had selected every recipient, merchant and intermediary, and no single relationship carried the whole decision. Different people held different fragments of the picture; their judgements overlapped, tested and sometimes corrected one another.

The team's differences mattered. A teacher noticed one thing, a pharmacist another. A quiet person heard what someone more forceful missed. One region questioned what another took for granted.

The seven in Syria gave us those different ways of seeing. The revolving team in London refined what they learned and turned it into something that could operate. Our technology preserved the record.

But the software could not see for them.

Ryan and Ryoma, the engineers now carrying Amanacard's platform forward, have brought perspectives of their own. They inherited not just the code but the reason the code exists. Together we are living through the growing pains of deciding where the boundary should lie as artificial intelligence becomes capable of reviewing and improving its own outputs: what machines can safely do, what still requires human judgement and how to keep human responsibility intact as tasks pass between people and machines.

Deciding how much freedom to give machines has made me look again at the freedom I gave people.

The other side of freedom

I have never wanted to manage people minute by minute. My instinct is almost entirely the opposite: to set the purpose, ask people to take responsibility and let them get the job done. The people I have worked with best — Fede, Dr Hamza and Clive in London; Saad; Örvar; and the seven — have generally wanted the same combination of purpose, responsibility and room to act. I made myself accessible whenever they needed me and assumed that availability compensated for how often I was elsewhere. They were rarely out of my thoughts.

But being accessible is not the same as being present.

From the team's side, I may have seemed to appear most clearly when something was not going as well as it could — or when an unauthorised space had turned Amanacard into "Amana Card", a major brand violation capable of summoning both me and my eighty-two-year-old father into the team WhatsApp chat. The joke had a history: years and a great many lawyers' hours had gone into securing trade mark protection for Amanacard around the world.

For the more consequential problems, I would arrive carrying the accumulated concern of weeks, intervene intensely, ask hard questions and expect the problem to be fixed. Then I would return to the other emergencies competing for attention. I thought I was giving people room. They may sometimes have experienced freedom interrupted by criticism.

Some of our software engineers have described our leadership as micromanaging. The word initially astonished me because I believed I was doing almost none of the management in between. I understand it better now. Intervention can feel like micromanagement when it arrives without enough of the companionship that should have preceded it.

I am apologetic for the way those intense interventions sometimes landed. I am not apologetic about performance in its plainest sense: whether Amanacard worked when somebody needed it. We were not building a forgiving product in a forgiving place. A missed detail could delay thousands of salaries, expose a hospital worker or lose the contract that paid our own team.

Behind any one of my hard questions might lie weeks spent finding a lawful route to pay those salaries, or resisting corruption that could have cost us the work altogether.

The friction was relentless. Working across languages was one of our strengths, but it could also turn a one-hour meeting into three. Basic applications were blocked. A conference call became an operation in itself: one platform unavailable, a virtual private network (VPN) failing, or a single-sign-on service dependent on a company that did not serve the country where a colleague lived. I often found myself acting as an IT operations manager before anyone could begin the job we were there to do.

Returning to Syria after the fall of the regime in December 2024, I was astonished by how little ordinary technology worked. My Adobe signature failed, delaying client communications. My phone could not perform basic functions. Services treated elsewhere as utilities stopped at the border. Yet Amanacard, built there at the height of conflict and cyberattacks, worked seamlessly. I marvelled all over again at what the team had achieved. Reliability was not a technical nicety. To the person at the far end, all our internal difficulties became only a payment that arrived or did not.

People deserve patience, context and recognition. They also deserve colleagues and systems that perform when the cost of failure will be carried by somebody else.

Fewer than two in five UK businesses survive five years. By the time I wrote this, Amanacard had survived ten, thanks in part to investors still waiting for a return that none of us expected to take so long. That statistic says nothing about the conditions of survival — whether customers pay, staff are safe or systems can fail without putting anyone in danger. Add responsibility for colleagues working under bombardment, and for their families, and the ordinary strain of keeping a young company alive becomes something else. Survival itself was an achievement I had rarely stopped long enough to recognise.

The recent transition in Syria made something else harder to ignore: the cost of my absences. One of the original seven — always friendly and funny, working in the region known as Euphrates Shield — has already moved on to manage a restaurant in Damascus. Saad had dealt with the change while I was consumed elsewhere, and I learned about it after the fact.

Managing the restaurant may be exactly the right next chapter for him. But I was unsettled by how quietly a person who had occupied so much space in my mind had moved out of the daily life of the team. I had assumed there would be time to return properly to the relationship.

My assumption that there would always be time followed me beyond work. For the better part of a decade, my friends got the fraction of me that survived the week, and some got less than that. Let me be exact about this. It was not indifference. It was Saturday morning, again.

Thinking about someone is not the same as making sure they know they matter.

Vienna

During those years, on a walk, Billy Joel's *Vienna* — a song I knew well — stopped being background. Worn thinner than I had realised by the gap between what we could do and what the institutions around us would allow, I heard each line differently. It was not asking me to abandon urgency; the fires were real. Nor was it asking me to become satisfied. I was not satisfied. I am still not. It offered something stranger: the possibility that meaning need not arrive on demand, and that slowing down might be how I lived long enough to hold the two photographs from Cannes together — the cost of living at the speed of the work, and the pride of knowing what the work had made possible. I still do not know what my Vienna is. But I can stop without becoming still, and return to the work anyway.

Making sure people know they matter is one lesson I owe this team. The other is how much people can build when nobody is expected to be the whole answer. Saad assembled the seven; the seven made the field visible; the London team turned what they learned into an operation; the engineers made it repeatable; and contractors, partners and families carried it through interruptions that no ordinary workplace would have asked them to absorb. Everyone held a different part of the work, and each part mattered.

What emerged was never the achievement of one person, one office or even the original seven. Amanacard was built across languages, borders, professions and years, by people noticing different things and trusting one another enough to bring those differences together. I remain amazed by what they made possible — and deeply fortunate to have made it with them.

6

THE RECORD-KEEPERS

In Peshawar's money bazaar there was a dealer who had worked the same spot in Chowk Yadgar for fifteen years, and each evening he performed a ritual I never tired of watching. Through the day, every transaction went into one big book — the day's churn in a single running column. After the gates closed, he moved each entry to its permanent place: regular clients into their own books, settlements squared, balances carried. A bookkeeping discipline older than the banks that would later call his trade untraceable. The received wisdom in London and Washington was that *hawala* was paperless. I sat in those shops for nine months and never once saw money move without being written down.

The books were closed to outsiders, which is a different thing entirely. That was the bargain of 2005: they let me read. It took Shah Wali's initial introduction, months of tea and returning, and the market's slow verdict that I was safe. In the end, the record-keepers opened their records, and everything I have built since rests on what they showed me. What changed in the years that followed was not that records began to exist. It was that some of them began to be kept in a form the outside world could read too. This chapter is about how that happened — and it begins in the fortnight when the Afghan state collapsed.

August 2021

When Kabul fell, the money stopped almost instantaneously. Western programmes closed overnight, leaving staff, drivers, suppliers, guards and teachers with no functioning formal channel through which to receive money due to them. The banks were imploding. The formal system's answer was the one it gives when risk becomes difficult to describe: too hard, too dangerous, wait. I did not think the people who had stayed behind should have to accept that answer, and by then I ran a system built for exactly this. So I went back to the market I already knew — no longer as a researcher asking to read the books, but as an operator asking to be written into them.

Doors opened quickly.

I knew Afghanistan province by province, not as names and boundaries on a map but as places I had travelled through, worked in and slept in. I knew the roads between them, the settlement practices, the *saraf* (merchant) commissions and the seasonal rhythms of liquidity. More importantly, some of the people involved remembered that I had come before to understand their trade rather than condemn it. My purpose was equally legible: people should receive what they were owed. That sentiment needed no translation.

Around me gathered an unlikely alliance: senior figures from the former republic's security establishment, now scattered outside the country, who wanted a way to help the people who had stayed. One of them had spent his career at senior rank inside those institutions. Chance had placed him abroad in the week everything fell, and it was too dangerous for him to return. His exile began as an accident of timing. He came instead to my flat in London, and at my kitchen table we spread out the registry of Afghanistan's licensed money dealers: more than three thousand names.

That number of names is an abstraction until it is paper on

a kitchen table. A database could tell us whether a name appeared on a sanctions list or another official register. It could not tell us whom to trust with other people's money. For that, we went through the registry one name at a time, each of us reaching into a different half of twenty years and consulting others who knew what we did not.

A 'no' could mean many things. A network that touched the opium economy. A cousin in an organisation we could not go near. A rumour neither of us could kill — and by then I knew exactly how much weight a rumour deserves, which is some, and never all. A 'yes' had to survive something harder: what we knew independently of the dealer, his corridors, his settlements, the company he kept and the company he refused. The formal screens still had to be passed. The real selection required memory.

From more than three thousand names, we arrived at ten whose selection we believed we could justify. We began with three. Every regulated institution on earth would call what happened at that table unscalable, and for years I heard that as an objection. I no longer do. Three thousand to ten is not a bottleneck. It is the control — the difference between a list that has been screened and a list that is known.

The reversal

Then came the conversation that inverted 2005. Sixteen years earlier, I had asked the record-keepers to let me read. Now I was asking them to let me write — to accept that beside their own books, every payment they made for us would be captured in a system of ours: a Quick Response (QR) code presented on a phone or card, an identity verified at the point of payment, and a record synchronised to a ledger a donor in Washington or a bank in London could inspect.

A virtual Amanacard, in a country where the previous government had just evaporated and the new one regarded almost everything foreign with suspicion. I was asking men whose survival strategy for four decades had been discretion to adopt, in the most uncertain month of their lives, a technology whose purpose was to make their work visible.

They said yes. Not because of the technology, and not simply because I had done the hard yards. Those years had taught me whom to bring into the room.

Few people knew the real name of the man beside me, but his presence made this particular request believable. He was serious, trusted and not someone who would attach himself to a proposition that might expose or deceive the merchants.

The QR code was simply the newest thing that trust was willing to carry.

We moved the stranded funds of the closed programmes through those first three merchant networks. Severance owed to Afghan staff and suppliers reached people displaced across all thirty-four provinces and beyond Afghanistan's borders, with a full audit trail behind each payment.

I cannot honestly give you the first payment as a single remembered scene; the opening weeks were a blur of them. So let me give you what each one looked like, because the invention was in the ordinariness. A worker owed months of severance stood at a merchant's counter in a market that had survived the collapse of two governments. On the phone in his hand, or on the card he carried, was a QR code. The merchant scanned it. The system asked the only question that had ever mattered in that market — is this the right person? — and a named human being answered it, on the record. Then the cash was counted out, which took as long as it always had.

For the worker, it was wages. For the merchant, another entry beside the book his trade had always kept. Far away, on a ledger the donor could read in real time, a line appeared recording what had happened, where, to whom and when.

The record then began to work in ways I had not designed. For some of the people we paid, our verification trail became the only credible proof of employment they still possessed — the reference their vanished employer could no longer write. The work spread from severance payments to education for women and girls, community health workers and Afghan cricket. Across seventeen funding rounds in one year, we made 5,341 payments totalling just over US$2 million to more than 1,800 players, staff and suppliers across thirty-four provinces. A national-team match fee, a groundsman's wage and a vendor's invoice could be paid in the same round, each leaving a record.

What a record costs

An Amanacard looks unremarkable: a small rectangle with no physical bank branch behind it and no slogan on it. The one promise it carries is printed in the holder's own language — that their data will be protected as though they stood inside the European Union. We made that decision at the very beginning, when the GDPR rules were barely in force — clients called it excessive, and lawyers across the sector called it unnecessary. Our lawyers did something better: they translated what we wanted into proper data protection practice. It seemed to us the minimum owed to people whose names could be dangerous to hold.

Because every card is a record — of a name, an entitlement, a payment that cannot afterwards be denied — and where we worked, records were never free. Someone had to make them, move them, explain them and keep them, and each of those acts carried risk. A record also cuts both ways: the card that proves to a committee in Berlin that a teacher was paid can suggest something quite different to the men at a checkpoint. What does it cost to keep a record where keeping records is dangerous? I had seen the answer in other people long before I had to carry it myself.

Dr Hamza knew the cost from the other side of the fence. Before he joined us, he was the manager who kept Al Quds hospital in Aleppo running under siege — collecting bundles of cash from an agent, hiding them in his coat pockets and glove box, and driving seventy-five kilometres past robbers and checkpoints, because without those journeys more than a hundred and fifty staff would have gone unpaid and three thousand patients a month would have been left untreated.

The revelation came later, from our office at Somerset House, when he saw the machinery from the other side: how the traditional aid channels had actually handled the field's information, and how differently it could be done. It was not the danger of carrying the money that had unsettled him, he has said publicly since. It was discovering what had happened to his identity along the way: to collect official aid he had handed over his personal details and had his ID and signature photographed on an agent's device, to be passed back through informal networks to organisations he would never meet. "I can only imagine how many people have my personal details."

The paradox is the point. The system that visibly controlled his data looked more intrusive — and protected him better — than the informality everyone assumed was harmless. When a man like that chooses to spend the next phase of his working life on the record-keepers' side, it tells you what the records are worth.

That lesson had not come only from data protection. It had come, more painfully, from cash itself.

In Syria, we had set a liquidity limit for each trusted merchant. Our operating procedures failed to hold one of them, leaving him to carry the equivalent of US$170,911 by road.

The journey went badly wrong. He was injured in a car accident, and the money was stolen.

The whole team mobilised. Two colleagues spent long days at the police station pressing for action, and sat with the community to talk through what had happened. Nine days later, an anonymous tip led police to a field, where they recovered US$90,000.

The recovery was captured on video. Loose banknotes lie scattered across the dusty ground, some caught in the scrub, others lifting and skittering in the wind, while a policeman in white scene-of-crime coveralls digs for more. It looked less like evidence from one of our payment rounds than a scene from a film.

I spent hours reconstructing how the cash had moved,

what had been lost and what could still be recovered, before taking the forensic accountant appointed by the charity through my findings.

The merchant accepted responsibility. He and his family surrendered another US$42,000 from their own savings. Within days, US$132,000 had reached the beneficiaries waiting for it; the remaining balance became a signed repayment plan.

On paper, the incident could now be reconciled: cash recovered, beneficiaries paid, balance acknowledged.

But it was happening to a man.

He had been injured, robbed and then left carrying a debt no household should have had to absorb in a collapsing economy. His family had emptied their savings to help make good a theft committed against him. Eventually, the weight became unbearable. He became suicidal and was hospitalised. Our field team visited him there and stayed close through his recovery.

That was why the limits mattered. They were not merely financial controls. They were a form of care. A payment can be reconciled while the person who made it possible is breaking underneath it.

Carrying the cards

I carried thousands of Amanacards into Afghanistan personally, trip after trip, and negotiated each arrival alone. I never paid a bribe.

There were tense encounters with Taliban soldiers at the border. I had to explain what an Amanacard was while deciding, very quickly, which part of the story might make sense to the men in front of me. The most effective answer was usually the simplest: teachers in villages whose names they knew would finally be paid, using UN money that had never reached them because those districts had been treated as Taliban territory and excluded.

Once, I was the only passenger left in the airport. I had been escorted into a room thick with cigarette smoke, where men came and went without explaining who they were or what they wanted. I sat alone with the cards and my documents, negotiating in fragments, watching the door and trying not to be transported back to the escape from Jalalabad I would prefer not to remember.

Eventually, I decided that waiting passively was making the room worse. I opened YouTube and found a clip narrated by my friend David Gower, the former England cricket captain with the unmistakable voice. It showed me walking and talking with Rashid Khan, Afghanistan's legendary spin bowler, who had agreed to become an Amanacard ambassador and support an appeal for a remarkable school in Kabul.

At first, the men watched the woman without a headscarf on the screen, not quite registering that she was me. They certainly recognised Rashid.

The phone began moving around the room. They replayed the footage, discussed his bowling and asked, with mounting disbelief, whether I actually knew him. I showed them a photograph of me wearing an Afghanistan cricket shirt, followed by pictures of our team at cricket grounds from Zabul to Paktika, handing out Amanacards and salaries.

The interrogation dissolved into cricket. Soon afterwards, my documents were returned and I was allowed to leave. I walked out confidently enough, though not entirely steadily.

The campaign itself had failed, even with such a prominent ambassador. We had aimed to raise US$256,320 to support the families of the school's students and staff through six months of winter. We came nowhere close. Without one businessman's extraordinary donation of US$20,000, we would not have been able to help even as many families as we did.

Cricket was not the only unlikely opening. On my first visit to Afghanistan's intelligence directorate after the takeover, the rooms still displayed challenge coins — the small commemorative tokens exchanged between military and government counterparts — and other memorabilia from the former government's relationships with NATO, but new men occupied the chairs beneath them. The contrast was surreal.

I had walked past rose gardens on the way in. Facing one of the scariest men in the room, I remembered that I had recently been in France and pulled up photographs of my mother's roses. He studied them. The awkwardness lifted enough for us to begin.

Cricket and roses created enough space for a conversation. I think the evidence has done the rest.

In time, we formalised what persistence had begun: an Afghan-registered, wholly British-owned entity, permitted to track international aid. It took seventy signatures, twelve months and no illicit payments — only repeated negotiation, some incredibly supportive people, and the accumulating proof that aid was visibly reaching districts that had received almost nothing for decades.

In the end, it was that evidence, rather than compromise, that secured the new authorities' consent to our operation. But the consent is not permanent. Officials change, permissions can be withdrawn, and evidence that persuaded one room may have to be proved all over again in the next.

The iPhone in the bazaar

The field corrects everyone.

We had designed our applications Android-first. It seemed obvious, grounded in experience: iPhones had been beyond the reach of most people we served, American products carried political baggage in many of the places where we worked, and Apple's closed ecosystem was difficult to enter. Chinese technology was cheaper and easier to work with.

Then I returned to the Afghan money markets in 2022 and 2023 and found iPhones everywhere.

In the two decades since I had watched merchants run a national financial system on push-button mobiles and handwritten ledgers, the iPhone had become one of the prizes of the American withdrawal — coveted, traded and carried through the bazaar like the flag of a departed empire.

We fought our way through Apple's approval process faster than I had feared. Another lesson followed: ground truth has a shelf life.

What the machine actually is

People who ask about the technology are usually asking

the wrong question. The machine is mostly method, and the method was forged in Syria: networks accredited one agent at a time, each held to the geography in which it was genuinely known; the points at which error or fraud could enter made visible; local observations recorded in the language in which they were made; and a named person responsible for every consequential judgement.

The software does not manufacture trust. It keeps the record beside the old one, so that a decision made locally can be inspected elsewhere without pretending it was made elsewhere. It is the big book in Chowk Yadgar, kept jointly now, with the outside world finally allowed to read.

None of it was built by choosing one side. We called the design our triple-A strategy: atomisation with centralisation, anonymisation with transparency, and automation with human intelligence. All three were governed by the same discipline: agility without sacrificing rigour. Like everything else in this story, each pair belonged on two sides of one coin.

In practice, it meant protecting identity at the edge while preserving controlled transparency at the centre. In the field, personal data remained encrypted and a device exposed only pseudonymised information about the people in front of it — nothing visible or stored in usable form amounted to a list. At the centre, an authorised organisation could link a transaction back to a verified identity within its own remit, with the full trail preserved for audit. The boundary between edge and centre was explicit, and crossing it required authorisation and was logged: a record of who had looked at the record.

The money was designed around the same principle as the data. Rather than concentrating it at the edge in one exposed sum, we divided it into smaller, less vulnerable allocations. The centre retained the consolidated picture and could reroute flows around any part of the network compromised by surveillance, robbery or hostile attention — anything that might lead harmful actors back to the people operating it.

On a whiteboard, refusing to collapse these tensions into a choice between opposites can look like indecision. In our world, it was the design, for the same reason the card carried its promise: every choice about a record is a choice about who can be hurt by it.

The method also has limits. It cannot conjure liquidity where a market has none, or replace the person who walks into a shop and notices that something does not fit. Its strength is the depth of particular human relationships, and scale has a habit of thinning exactly that. The task is to extend the reach of those relationships without diluting their substance.

That distinction matters more now. A large language model can absorb a record no colleague could reasonably be asked to hold, and it can produce the feeling of a conversation around that record. It can help a person remember, compare and ask better questions. But relationship is not just context at scale. It is the history of having shown up, been tested, disappointed someone and repaired it, kept a confidence, refused a shortcut, carried a consequence. AI may scale the experience of being known. It does not, by itself, create the mutual obligation that makes knowledge safe to hold.

Full circle at the Intercontinental Hotel

There is a version of this story in which it is all new. But I know precisely where the old version is buried.

In 2011, USAID, then America's aid agency, convened a mobile-money summit at the Intercontinental in Kabul, part of a programme worth hundreds of millions of dollars to modernise Afghan finance. I was the only woman on a panel before senior officials and bank and telecommunications executives, several of whom made no secret of what they thought of my being there. I reported what the fieldwork said: the future of Afghan payments would be built through the trust networks that already worked, or it would not be built at all.

It was not what the people controlling the aid money wanted to hear.

Two memories remain from that day. At the back of the room stood one of my oldest friends, then serving with the Australian Special Air Service (SAS) and looking uncannily like the figure in Ian Coate's poem: moving "freely in the daytime, but always out of sight", and appearing where least expected. As the panel bristled, an enormous grin began to escape through a usually concealing beard.

Afterwards, he told me what he had seen from there. The moment I finished speaking, the Afghans in the room began to gather around me — not the international contractors and consultants, but the people who knew the market. Among them was a deputy governor of the central bank, who asked the question the modernisation initiative had not answered:

How do you get them to trust you?

The programme spent its hundreds of millions, yet even before the fall fewer than one Afghan in ten had access to a bank account. A decade later, a tiny team with QR codes and ten carefully chosen names went where all that money had not. We had reached every province, even the remotest villages.

After the banking system collapsed in 2021, one US contractor that had spent six months trying unsuccessfully to move money into its Kabul account used our system to pay more than US$1 million in outstanding obligations. What began as rescue work became routine: US$4 million in health payments reached workers across nineteen provinces; another US$7 million in education payments moved across seventeen.

I could watch the verification happen in the record: the recipient using a verified card, entering their PIN twice, and confirming in real time that the money owed to them had been received. There was a person at the other end of the transaction, showing us that the promise had been kept.

The numbers mattered, but I remember the faces. In photographs sent back from some of those payments, people looked into the camera after learning that what was owed to them had finally arrived.

We did not try to replace the record-keepers. We asked them to let us into the ledger.

The answer

So: the deputy governor's question, fifteen years late. How do you get them to trust you?

You go to every province, and you go back. You gladly eat what is put in front of you and learn what the ledgers mean before asking to change them. You want something they can verify — people receiving what they are owed — and you keep wanting it when it stops being convenient.

You do not pay the bribe, because once the first bribe enters, the record no longer says what it appears to say. Then you ask for less than everyone before you asked for: not that they abandon their books, but that they let you keep one beside theirs, written so that both worlds can read it.

That is what a record costs where records are dangerous: someone decides, every day, that it is worth carrying — and decides, just as deliberately, what it will never say.

Trust is not a technology and it is not a policy. It is the compound interest on kept promises.

The record-keepers know the calculation. They can read your balance at a glance.

7

THE THEATRE OF COMPLIANCE

The most instructive interview I conducted while researching the remaking of Afghanistan's banking system took place nowhere near Kabul. It was in a hotel in downtown Dubai — marble, glass, air conditioning pitched at Swiss-alpine — where I sat across from Sher Khan Farnood: the largest *hawala* operator Afghanistan had produced, a World Series of Poker champion and, by then, founder of the newly licensed Kabul Bank, which would become the country's largest private bank.

Everything about the meeting was an exercise in translation. He had exchanged the bazaar for a boardroom, client books for a core banking system, and the reputation of a dealer for the letterhead of a bank. He wore the new costume well, in the way a poker champion wears anything: as a choice, fully in his control, revealing nothing he had not decided to reveal.

We talked for a long time about the old trade and the new one. The longer we talked, the more clearly I could hear that only the vocabulary had changed. So, towards the end, I asked him directly whether he had simply given his *hawala* business a kind of figurative face-lift.

He smiled.

"Yes indeed."

No hesitation, no offence taken — the pleasure of a card player showing you, just once and because it cost him nothing, exactly what he was holding. It was the most honest sentence anyone in a suit said to me that decade, delivered in a hotel designed to make such sentences unnecessary.

That exchange became the seed of a concept that ran through the final chapters of my first book: *regulatory ritualism*

— choosing actions that look good to the regulator over actions that make a difference. Sher Khan had learned to speak fluent compliance. The vocabulary was Basel, but the discipline of the bazaar — where reputation was collateral and failure carried consequences — had fallen away. He no longer needed to be trustworthy, only to look regulated.

The cost became clear years later when Kabul Bank collapsed in a fraud so large, relative to the economy beneath it, that it shook the Western project in Afghanistan. A man who had mastered the language of banking regulation had presided over precisely the kind of banking failure that regulation was meant to prevent.

The original repertoire

The theatre went far beyond Sher Khan. During the fieldwork I watched the original repertoire assemble. There was the Money Exchange Dealers' Association, presented earnestly to visiting Western officials as the market's self-governing body. The dealers themselves told me, with some amusement, that it had no real power; adjudication still ran through respected elders.

There was the central bank's new licensing regime, which many dealers treated as a price to be paid rather than a discipline to be absorbed. And there was the memorable candour of a senior anti-money-laundering official in the Gulf, who told me that his unit did not expect suspicious-transaction reports from *hawaladars* at all: the system is built on trust.

Many of the people closest to the performance knew exactly

what it was. The people most likely to mistake it for the work were the distant audiences who had commissioned and funded it.

In 2006 I adapted a term from research practice to describe this: *optimal ignorance* — the deliberate avoidance of uncomfortable facts that might impede the momentum of a project. I have spent the years since watching the West stage its own, far more expensive production of the same show.

Our repertoire

What passed for due diligence in one of the deadliest corridors of the 2010s was the familiar three-quote tender, followed — where screening happened at all — by a rough Latin-script rendering of an agent's name and perhaps an unverified date of birth run against a sanctions list and filed as counter-terrorism.

The forms were completed. The boxes were ticked. The files would have survived any audit designed to establish that files existed. Meanwhile, the questions that mattered — who controlled the network, where its liquidity pooled, whose war it might be funding — went unasked because no form required them. That is ritualism with a Western accent: the bazaar performs for the regulator; we perform for ourselves.

The performance has a constituency, and I learned its size the hard way. Our methods in Syria — a small field team, specialist domain knowledge and technology built around their judgement — quietly threatened aid operations employing hundreds of people because we kept demonstrating that many of the layers between donor and recipient were removable.

Layers defend themselves.

The most candid explanation came from a senior NGO representative at a conference, on the record and with permission to repeat it: "We are not being told by donors to change the way we are working, so why would we do anything different than the current manual methods of dealing with *hawala*?"

Not long afterwards, a partner from a leading London audit firm overseeing some of the largest multilateral aid programmes expressed much the same preference, for much the same reason. His government clients had not asked the firm to validate a new system. They had asked it to audit the existing one. He would therefore rather work through thousands of paper records in a language his team could barely decipher than rely on clean, real-time transaction data that could be analysed in minutes.

No invented villain in the story ever says anything so clarifying. The theatre continues because the audience keeps commissioning it.

Yet none of these people are monsters, and this book does not deal in caricature. The compliance officer filing the ritual screen is doing what the institution measures and pays him to do. The aid official defending a hundred-person operation is protecting colleagues, careers and a mission she believes in. The auditor is answering the question the government client has paid him to answer. The donor renewing the manual method is choosing a failure he knows how to administer over a success he would have to explain.

That is precisely what turns the theatre into *The System*

rather than a scandal: everyone can blame it, and no one has to answer for keeping it intact. Exposing one person changes little when the rewards around them do not.

While finishing this book, I heard the historian Ian Kumekawa describe what passed for a defence of the offshore world as a shrug. The shrug says it all and answers for nothing.

The irony sharpened in August 2026, when FinCEN permanently exempted US-created companies and US persons from the federal regime for disclosing who ultimately owned or controlled a company, leaving only certain foreign companies within its scope. Years in the making, the requirement had been hollowed out before it had properly taken hold.

Inside the machinery

Recognition for trying to do things differently could carry constraints of its own, drawing us further inside the very system we hoped to change. We discovered that in 2018, when a Western government aid department selected Amanacard for a prestigious six-figure technology-innovation award. We were small, unprofitable and permanently short of money. Yet after months of scoping, and before there was a final contract, we withdrew. The more we learnt about what accepting it would entail, the less the award resembled freedom to build: meetings about the work rather than time to do it, and the expectation that some of the money would pay for support assigned to us rather than chosen by us.

One humanitarian adviser inside the department had championed us with unusual persistence, and I was genuinely sorry to disappoint him. He was not *The System*. Nor were the advocates I later encountered inside other European institutions, who tried just as hard and were equally unable to make their machinery move.

That was the lesson: institutions do not move by themselves. People inside them must push — and sometimes even the most determined advocates cannot shift the machinery.

We stayed poorer and less beholden, but not free. More than once, we considered saying plainly in public what our ledgers and experience showed us in private. Each time, the arithmetic was brutal. We knew by then that vindication could take years, and that the person vindicated might still be pushed out of the room where decisions were made. So we chose the slower course: to stay inside, say less than we knew, and return to the work the next morning.

Fairness demands the other side of that account. Sometimes the people inside an institution pushed hard enough to make it move. At Herbert Smith Freehills Kramer LLP, one of the world's leading international law firms, Dan, David and Jeremy did. They proved that not every meeting between our world and the rulebooks ended in performance.

The firm has given us years of first-rate legal support, pro bono — committing extraordinary time to problems that may have been commercially small but were operationally enormous to us. We brought the field knowledge and niche expertise. We identified the risks and wrote them into a register so comprehensive that our first institutional investors, from a major development bank, looked up from it wide-eyed. HSF Kramer worked through the register with us, helping us decide how each risk could be mitigated and building the resulting protections into our policies and contracts.

The clearest expression of that work was in the force-majeure provisions — the clauses that deal with events so exceptional that ordinary contractual promises can no longer be kept. In most contracts, force majeure sits at the edge of possibility: war, catastrophe, the breakdown of systems on which the agreement depends. In ours, it occupied much of the map. We were designing in advance for conditions serious enough to deter most organisations from entering such territory at all. In the years since, every risk in that register has come at us. We were not spared; we were ready.

That is the alternative to theatre: rules meeting reality and adapting to it, rather than being performed for an audience elsewhere. Institutions can adapt. They simply do so less often than the stakes warrant.

The market that does not want to see

The problem was hardly obscure. The US government's own inspector general had documented for years that funds intended for Afghans leaked to the Taliban through taxation, fees and diversion. Congress demanded that it stop. Donors spoke constantly of transparency and accountability.

We built an instrument that turned those abstractions into evidence. It made the journey of a payment visible from source to recipient, and we proved it in the corridors considered hardest to reach. Yet the aid system continued to move much of its money through *hawala* without naming the intermediary, or through banks that quietly relied on the same dealers we accredited openly. The bank's letterhead made the route look formal. It did not necessarily change who carried the cash.

Transparency and accountability are easiest to praise while they remain principles. Once translated into a transaction record, they become less comfortable. The record shows where the money went and whose choices shaped the route. That leaves less room for responsibility to be passed down the chain — from donor to contractor, contractor to bank, bank to policy — with the payment's actual journey visible.

When Britain discovered in 2023 that a 2022 Afghan resettlement data breach had exposed the identities of thousands of people, the danger was not merely administrative. Many of those affected were families Britain had promised to relocate, but who remained stranded as resettlement became politically and operationally difficult. The breach remained secret until 15 July 2025, when Defence Secretary John Healey disclosed it to Parliament.

Even before the breach, there was an unresolved practical question: if relocation could not happen quickly, how could those families be supported in the meantime? And if financial assistance or compensation were offered, how could it reach them discreetly, without creating another trail that placed them at greater risk? Once their identities had been exposed, that question became more urgent.

By then, we had a proven, licensed and independently audited route for doing exactly that: verifying recipients discreetly, delivering cash quickly and at scale across

Afghanistan, and limiting who could see their personal information. On 17 July, two days after the parliamentary disclosure, I wrote directly to the Defence Secretary and offered it.

The reply arrived two months later from a secretariat. It thanked me for my valued perspective, assured me that robust data-handling practices were now in place, and supplied two links: a public hotline for affected people to telephone and the login page for the Defence Sourcing Portal.

Perhaps the second link was an oblique invitation to register as a defence contractor, or simply a courteous way of sending the correspondence elsewhere. What the reply did not do was engage with the proposition in front of it: that a safe route already existed for reaching people whom Britain had undertaken to protect.

The response offered procedure where the situation required judgement — another instance of a recognised sign of safety obscuring the risk that actually mattered.

The seatbelt

I had learned that lesson more literally in Baghdad, at speed, to the flash of a sparkling phone case. A fintech client had sent me to conduct due diligence on a local bank before entering the Iraqi market, and one of the bank's employees adopted me on sight. She became my guide, my translator and — most alarmingly — my driver.

She was unforgettable. The car had Minnie Mouse seat covers. She dressed as if heading to a gala — perfect hair, immaculate make-up — and took selfies while driving, which she did fast. Lanes appeared to be advisory. Pavements occasionally joined the road. Roundabouts in central Baghdad were not so much navigated as attacked, with bursts of acceleration, sudden swerves and the occasional one-handed selfie mid-rotation. Every instinct I possessed as an Australian screamed that I was about to die.

Then, quite casually, she reached across, pressed the button on my seatbelt and let it spring free.

I stared at her.

"It isn't safe," she said.

I looked at the empty belt in disbelief. She nodded towards the window.

"They'll know there's a foreigner in the car."

The seatbelt was not protecting me. It was identifying me.

By the end of the week, I had a matching lesson to file professionally. The person most outsiders would have identified as the unsafe part of the journey had spent it managing risks I did not know existed. The institution everyone assumed was the safe part required a harder conversation: my report concluded that its controls were not yet strong enough for what my client wanted to build. It was not the conclusion the operations team at headquarters had hoped to receive.

The scene is this chapter in miniature. The bank had marble, policies, committees and the reassuring vocabulary of compliance. Local operators above their shops, with battered desks and handwritten ledgers, often possessed something harder to manufacture: detailed knowledge of the people they served — their families, businesses, ordinary transaction patterns and the instinctive recognition of when something did not fit.

That does not make local operators inherently better than banks. It means trustworthiness cannot be inferred from appearances.

I never wore a seatbelt again in a conflict zone.

The whistleblower's ledger

The theatre extends even to the instruments intended to expose it. The evidence is on the public record. In 2007, the UN hired me to evaluate a major programme in Somalia: an initiative, well ahead of its time, intended to bring *hawaladars* together around common technology. My findings were not what the country office wanted. A second consultant's report appeared. I was pressed to endorse it in place of my own. When I declined, a 'consolidated' report was issued over my name anyway, while my fee was quietly blocked month after month.

I still have my reply because it later surfaced in the press: 'I would love to comply with your request ... if for no other reason than to expedite payment for my fee. I hope however that you'll appreciate as a matter of principle that I am unable to do that.'

I was young and unpaid, and I have never regretted it.

I was not the only person who saw what was happening. One of the programme's key technical experts, a Somali named Ismail Ahmed, had been watching from inside: fraudulent payments, bogus contracts and the institutional pressure not to make them apparent. He was also that rare colleague who respected thoroughness at a time when it was usually treated as an inconvenience. We spent days in Somalia discussing what the remittance industry could become.

Then he blew the whistle. He was punished, took the UN to its own Ethics Committee and won: the ruling found that he had suffered retaliation for protected disclosures. My substituted report and blocked fee appeared in the coverage as part of the same pattern, with my protest emails quoted verbatim — disclosed by Ismail as part of his paper trail.

The official at the centre of the episode was not punished. He was promoted to run the UN's development operation in Haiti. Days after the 2010 earthquake, standing amidst the chaos that was Port-au-Prince, I looked up and saw the same familiar figure walking past in his suave, unhurried way. For a moment I thought I had seen a ghost.

The system had relocated the problem and called it a career path. The whistleblower and the evaluator had paid for what they recorded.

But every ledger has two columns.

Three years later, I was standing on an airstrip in the Philippines — the typhoon deployment of Chapter Four — when my phone rang with a London number. An analyst at Accel Partners, where I had been working on the morning of September 11, was conducting diligence on a Somali entrepreneur and needed a character reference.

Ismail Ahmed.

I had no idea what he had been building. I knew exactly who he was.

I stood on the runway with the wreckage of Tacloban around me and described the record I knew: a man who had watched a system defraud some of the poorest people on earth and had risked his career to document it. He understood the remittance market, the problem and the scale of the opportunity. More importantly for the question I had been asked, he had shown what he did when telling an institution what it wanted to hear would have been easier.

He reportedly used part of the compensation from his retaliation case to help seed WorldRemit. Months later, Accel led a US$40 million investment in the company — one of the largest European Series A rounds of its time, in what polite finance regarded as the dirty end of the money business.

What had cost him his career in one institution had helped finance the company another was willing to trust with millions.

That is the other side of the ledger. A decision that looks ruinous in one moment may become part of how a person is understood years later. The contracts we declined and findings we would not soften often felt like uncomplicated losses at the time. Later, those decisions became part of why law enforcement officers kept speaking to me and why three money dealers in Kabul said yes to a stranger's QR codes in August 2021.

The map in the conference room

By June 2023, much of that accumulated record could be described in a room.

At a World Bank conference on Afghanistan payments in Dubai — the first time I spoke publicly about Amanacard's work there — I arrived carrying a large, beautifully printed map of the country. It showed nothing more sensitive than Afghanistan's geography: its provincial boundaries and place names.

Trust had featured in the discussion for two days. A colleague from the Aga Khan Foundation had even used the phrase *trust is the coin of the realm*, and I was able to point out, to some amusement, that I had written the book.

The Dubai hosts saw no difficulty in displaying the map during my talk. The organisers did. Because it had not been submitted and approved in advance, it somehow became a problem. I spent five minutes arguing for permission to display an ordinary map of the country named on the programme — an exchange that told me more about the institution than much of the preceding discussion.

When my own five minutes came, I used the map only as a point of orientation. For the pleasure of the alliteration, I chose places beginning with G: Gezab, where we had paid teachers; Ghor, where we had paid community health workers; Ghazni and Gardez, where we had paid cricketers and their families. There were no pins and no data marked on it. I placed them there verbally, telling brief stories about activity that had actually taken place in locations many people in the room would have struggled to reach.

People came up afterwards and said that the presentation had grounded the event in the place — in the country the conference was about. I had finished by asking the room this question:

How many really want to know what is going on — where the money is really ending up?

There was no answer, and none of the major stakeholders

followed up afterwards. I found the silence harder to understand than an objection would have been. The stories had made visible, for a few minutes, the thing years of strategies and conferences had claimed to want: actual payments to actual people in actual places on a map, with evidence behind them.

The pieces in the air

I no longer expect a single room to change this. Systems change when their standards change. That is why, in 2026, my group submitted formal evidence to FATF, the body that writes the world's financial-crime rules, as it prepared guidance for the revised *Recommendation 16* — the global rule governing the information that must travel with a payment.

We argued for outcomes over prescribed forms: define the assurance a check must deliver, then allow each corridor to prove it in a way its reality can support. Mandate a ritual and the world will perform it, whether or not it works.

The revised standard will shape cross-border payments into the next decade. Its guidance and implementation will reveal what the system really values: evidence that a payment is transparent, or the appearance of transparency.

The same question now hangs over aid, though the rupture has been far more abrupt. In 2025, Washington dismantled USAID as an operating agency, cancelled many of its programmes and transferred a much smaller residue to the US State Department. Some called it vandalism. Others, more quietly, had long said the system needed to be shaken. I can hold both thoughts at once.

I had seen the image before. "The kaleidoscope has been shaken," Tony Blair told his party conference weeks after the attacks of 2001. "The pieces are in flux. Soon they will settle again. Before they do, let us re-order this world around us." I opened a chapter of my first book with those words because they captured the confidence of that moment: the belief that order was something the orderly brought to other people's chaos.

Now the kaleidoscope has been shaken in Washington itself. The pieces of the aid system are in the air. Nobody knows what will settle in their place: a smaller system that finally rewards results, or the same theatre with fewer seats. The last mile will feel the answer first.

I will keep the ledger either way. The dealers' association with no power, the face-lifted bank, the three-quote tender, the procurement portal and the conference that applauded and went silent were not separate failures. They were the same play in different costumes: actions that looked good preferred to actions that made a difference.

The bazaar performed for the regulator. The rest of us built institutions that could no longer tell the performance from the work.

Language had become theatre too. Compliance, sustainability, resilience, inclusion, localisation: all necessary words, all capable of becoming costumes once they are separated from the obligations they name. A system can learn the right vocabulary and still fail the person in front of it. That is why the question cannot be whether a word appears in the strategy, policy or model output. The question is whether reality can still answer back.

8

THE BOOK THAT MUST NOT BURN

I want to scream. I found the sentence in one of my old notebooks from the Papua New Guinea Highlands, written in the hand of a much younger woman.

I did not remember writing it. I remembered the feeling.

The sentence could just as easily have appeared in my notebooks from eastern Democratic Republic of Congo, Myanmar, Sri Lanka, or Sudan. Most often it came from the same injustice: people punished by the accident of having been born in what the rest of the world treated as the wrong place.

The sentence sat among testimony from women who had survived things no one should have to describe twice. Yet the encounter that set my leg shaking beneath the table and my pen racing was with a Catholic priest in Port Moresby.

I asked him why the Church could not do more to help with tribal divisions. We had, after all, the same faith. The same belief in God.

He looked at me with something between pity and irritation and scoffed.

He said he sometimes wondered whether that was true.

It did nothing to rock my faith. It made me angry. People could pray, sing and preach, then still hurt one another.

Faith was not proved by the words. The test came when power, fear, money, obligation and shame entered the same room.

The wrong place

I left PNG in 2003 and went almost directly to Afghanistan.

On the face of it, the two countries had almost nothing

in common. One was a mountainous collection of tropical islands divided among more than eight hundred languages; the other was landlocked, shifting from fierce summer heat to deep winter snow, with the politics of its neighbours reaching into every valley. Most Afghans — and many of the internationals I met — had never heard of Papua New Guinea. When I tried the comparison aloud, it did not travel very far, and I soon stopped making it.

But privately, the feeling stayed with me.

I kept encountering the same kind of complexity: places where the state existed but did not reach evenly; where relationships carried more practical authority than institutions; where money, obligation, kinship and politics were difficult to separate; where outsiders mistook what they could not read for disorder; and where people had developed ways of getting things done that made perfect sense once somebody committed the time to understand them.

The violence travelled too. Its forms changed — tribal conflict, insurgency, child soldiers — but its architecture remained: weapons outrunning restraint, authority without consequence, and women carrying the aftermath.

Some countries could stop me in my tracks.

Afghanistan had lakes so blue they seemed to sting the eyes, and mountains that swallowed sound. Papua New Guinea had jungle, sea and a light so warm that the landscape looked newly made. Eastern Congo had lush green hills rising from black volcanic ground, as though the tropics had grown over the moon.

Beauty did not soften the difficulty. It made the violence more bewildering.

In Darfur, the desert and the camps for displaced people seemed to announce their harshness. In the Democratic Republic of Congo's North Kivu province, suffering hid beneath the green. Camps outside Goma stood on broken lava — black, jagged and almost impossible to level into a field where children could play. Even the tents, meant as shelter, became places of danger.

Women told me of armed men coming after dark and gang-raping them inside the very places to which they had fled for protection. One woman lifted her shirt to show me the raw scar where a knife had opened her torso. After it was stitched, she said, the same men returned the next night, jumped on the wound and raped her again.

At the back of a report I later wrote, *Principled Pragmatism*, one photograph held the problem in a single frame: a World Vision vehicle carrying therapeutic food to the Rwanguba Nutrition Centre, passing a rebel soldier on a road that had reopened that day after heavy fighting.

The photograph was not an argument about whether food or cash was the better response. I was not yet asking that question, and I cannot now say which would have worked there. It captured a more immediate dilemma: on that day, reaching the children meant taking aid through territory controlled by armed men.

That was what I meant by principled pragmatism: preserving humanitarian independence without pretending that the people controlling access could simply be avoided. Refusing all contact might keep the principle looking clean while leaving civilians beyond reach.

The photograph did not resolve the dilemma. It showed it.

The return

Two decades later, when I returned to Papua New Guinea, it did not feel simpler. It had never been simple to me. What had changed was that some things felt unexpectedly familiar.

Familiarity arrived through the body before the mind — the taste of fresh barramundi fish, fruit and vegetables; clothes chosen for the heat rather than, as in some of the places I had worked, for the scrutiny of a hundred watching eyes that seemed to follow a foreign woman; and a dry, teasing and slightly irreverent humour that often broke through at precisely the moment a room became too serious.

After years of calculating head coverings, sleeve lengths and whether the shape of my clothes might offend somebody, I felt the small freedom of simply getting dressed.

Faith, too, was spoken aloud. The priest's scoff had lost none of its force, but in Sunday services we at least addressed the same God.

While staying as the guest of an oil and gas company, I found a tiny bush church. Little English was spoken, but a Papua New Guinean colleague from another part of the country quietly translated, carrying surprisingly deep ideas between us. That morning the preacher spoke about the same Bible passage I had been reading, alongside a commentary on it, only the night before. When I told them, the congregation marvelled at the timing: a stranger from somewhere so far away had been studying the very same passage only hours before walking into their church.

An elderly woman beside me, her spectacles balanced on the tip of her nose, shared her Bible and a ruled schoolbook in which she appeared to have copied favourite verses over many years. The congregation sang, adding words to welcome me. The church could hardly have been more local, yet for that hour no explanation of distance was required.

The priest in Port Moresby had made me angry at what an institution could do while speaking the language of faith. Here in the Highlands there was barely an institution at all: a congregation, a Bible and an elderly woman's handwritten book. Yet the thing itself felt entirely present.

I thought later about a pattern that runs from the first pages of Genesis: God speaks creation into order, and breath gives life. Words can order a room, a liturgy, a policy or a database. They can separate, name, promise and record. But words alone do not make a thing alive. Life comes with breath: body, presence, attention, vulnerability, the costly fact of being there. That was not an abstract theological point to me. It was the difference between a statement of faith and a woman sharing her Bible; between a safeguarding policy and a person kept safe; between a record existing and a life being recognised by it. That is one limit no machine overcomes simply by finding a voice or acquiring a body. Lived creaturely experience is not the same thing as fluent performance.

Papua New Guinea remained daunting. Its violence was shocking, but no longer singular. Fraud seemed woven through too many transactions. Wealth was being drawn from the ground at enormous scale while ordinary people struggled to receive small amounts safely.

But as I began looking more closely at the extractive industries — many of whose decisions, capital and professional networks ran through Australia — the problems no longer seemed quite as exotic as I had once allowed them to seem.

I came, after all, from the Lucky Country — a phrase coined as a criticism and cheerfully adopted by Australians as a boast.

Australia knew how to extract value, secure it, finance it and move it. Australian companies, advisers and operators were deeply involved in the systems surrounding Papua New Guinea's resources. The separation between their difficulties and ours was therefore less comfortable than the map suggested. It was easy to describe violence, corruption or exclusion as features of somewhere else. It was harder when the money, management and assumptions often led back across the same sea.

The women brought that proximity into focus.

On my first visit to the Highlands, I was given two *bilums*. I took the woven string bags home and put them away, appreciating them as gifts without understanding their significance. It was only when I returned, more than two decades later, that I began to grasp what had been in my keeping all that time.

Women carried *bilums* from the head or shoulder, filled with food, possessions, sometimes a sleeping child. Each was practical and personal, always one of a kind. The knowledge of weaving them passed through generations of women; in Tok Pisin, *bilum* also means womb. A single bag could hold the daily life of a household, the history of a family and, for the woman who made and sold it, a measure of economic independence.

The *bilum* began to feel to me — not as a translation of *amana*, but as a material echo of it: something entrusted for safekeeping, held close to the body, carrying value and obligation at once. The bag itself was visible. The life and labour it carried could remain unseen by formal systems.

I had spent the intervening years trying to solve versions of that problem elsewhere. In Afghanistan, Iraq, Somalia, Syria, Yemen, the person at the end of a payment was often the least visible part of the chain. Institutions could account for the transfer leaving headquarters while knowing remarkably little about the hand that finally received it. We had built identity, payment and cash-tracking tools to place that person back into the record without stripping away choice.

Returning to Papua New Guinea, my question was not whether I could import a solution perfected somewhere else. Nothing had been perfected, and PNG was not a blank surface waiting for foreign design.

The question was whether anything I had learned across all those other places could be useful in a country that still resisted easy explanation.

The return also made me look differently at Australia. I had spent years building systems intended to keep a person visible without taking away the freedom that made the person themselves. I thought I understood the danger in the abstract. Then I read about a woman who had spent much of her life fighting to remain in possession of herself.

Truganini

After a short visit to Tasmania I read Cassandra Pybus's

account of *Truganini* — an Aboriginal Tasmanian woman. Not many books make me cry. That one did, more than once.

In 1829, Truganini met George Augustus Robinson, the English colonist appointed to "conciliate" Aboriginal Tasmanians and bring them into colonial settlements.

She understood that Robinson needed her. He depended on her knowledge of Country, her command of languages, her ability to navigate, find food and approach people he could never have reached without her. She went with him when she believed that doing so might protect her people from something worse.

But she never surrendered herself completely to his plan. At times, she simply disappeared to live as she chose, then returned.

Nothing about her circumstances was safe. But she was not powerless either. She knew that her knowledge made her valuable enough to be taken back. Within the appalling limits placed around her, she preserved some capacity to refuse.

She was a formidable force of nature.

That was why Pybus's description of her grotesquely swollen ankles made me recoil — not from Truganini, but from what had been done to her. Pybus attributes the swelling to venereal disease, probably syphilis.

The infection belonged to the same colonial world in which sealers, whalers and other men abducted, raped and sexually exploited Aboriginal women. It debilitated a young and otherwise extraordinarily capable woman: a powerful swimmer who crossed dangerous water and pulled other people through it, and whose strength and knowledge men like Robinson depended upon.

The scene that finally broke me came in a boat.

As I remember it, Truganini was being carried through waters she knew intimately when she fell to her knees and begged the Englishman beside her to promise that, after she died, he would place her body in a sack and throw it into the deepest part of the ocean.

I put the book down. By then, I knew enough to dread what would come next.

She knew what colonial institutions had done to Aboriginal bodies: measuring, classifying, collecting and displaying them. She feared that the men who had claimed authority over her people in life would claim her body in death.

The ocean was not surrender. It was the last place she thought they might not be able to own her.

She was wrong.

Two years after Truganini died, her body was exhumed. In 1904, her skeleton was placed on public display in the Tasmanian Museum. It remained there until 1947.

The very future she had begged to escape.

She had asked to be thrown into the deepest water. Instead, for more than four decades, strangers looked at her bones.

That knowledge changed the memory of something I had seen years earlier in London. I remembered an exhibition about Aboriginal Australia and British newspapers reporting on ancestral remains being returned from museum collections. At the time, I registered the issue as important but distant: a historical wrong being addressed by institutions somewhere above me.

Reading *Truganini* collapsed that distance.

These were not simply "remains". They were people who

had understood what might be done to them and had been denied even the power to keep possession of their own bodies.

Another part of my education came by chance, during a long delay on the tarmac in Melbourne. Beside me sat an elegant woman perhaps ten years my senior, working on a tapestry. Sam was a nurse whose Aboriginal heritage came through her mother's line. I admitted how little I knew and asked her to educate me.

She did.

That conversation, and the book, changed how I heard the words that opened Australian events and meetings. I had only recently returned after years away, and the practice was unfamiliar: a Welcome to Country, given by a Traditional Owner authorised to welcome visitors; or an Acknowledgement of Country.

The names of the peoples and nations were unfamiliar to me. The people saying them did not always fit the crude categories I had absorbed without noticing. The nurse had blonde hair, delicate features and deep brown eyes. History did not always announce itself in the way I expected.

The words should matter. But repetition could also turn invitation into wallpaper. Where I sat, they interrupted the proceedings without necessarily interrupting the assumptions or power inside the room.

The meeting continued.

None of this had been secret. Books existed. Aboriginal people had spent decades telling these stories. I had not looked hard enough.

Nor had school taught us much. I had grown up fluent in Aboriginal place names and almost ignorant of what they contained.

What's in a name?

Once I started looking, Aboriginal words were embedded in the map of my childhood.

Moorakyne Avenue was where one of my closest childhood friends lived. I spent countless days crossing the road to her house for play dates. Toorak was where I went to school. Prahran and Kooyong were where we bought groceries and visited other friends.

The names were effortless in my mouth because I had said them all my life. Yet I had no idea where most of them came from, whose language they carried, or what they had meant before they became suburbs, roads and railway stations.

Meanwhile, reading *Truganini,* I struggled to retain the names of some of the people around her. Their length and unfamiliar structure made them difficult for me to hold. I found myself going back over them and, sometimes, allowing my eyes to pass over them.

That troubled me. Truganini had just forced me to think about what remains of a person when other people take control of the record. A name seemed the smallest possible place to begin resisting that disappearance. Part of remembering a person is remembering their name.

The lapse took me back to the women I had watched in blue burqas recede against the sky until I caught myself registering only the category — *Afghan women* — and no longer the person within it. I felt something similar happening on the page. When I allowed an unfamiliar name to pass without holding it, the category remained and the person disappeared.

Complexity does not always announce itself as erasure. Sometimes it simply invites the hurried reader, official or database to move on.

A difficult name is more likely to be misspelt, broken into the wrong fields, rejected by a system or quietly dropped from the record. The institution calls it a data-quality problem. The person experiences disappearance.

We built our transliteration system by hand. Tens of thousands of names were reviewed with professional linguists and local colleagues who understood the differences between regional Arabic, Pashto and Dari, and the many versions of a single name created when families, banks, governments and aid agencies each wrote it differently.

I want to record the human work behind that system, particularly Kholoud's. From our London office, she reviewed the names of the first thousands of people to receive an Amanacard in Syria. Her tools were her own knowledge, a spreadsheet and long conversations with colleagues including Dr Hamza, Saad and Abdulkhalek, and later Firas and Adnan.

They debated spelling after spelling — not to impose one supposedly correct version, but to preserve the name while making it usable across systems that did not know what to do with its complexity.

It was painstaking work. It almost drove Örvar mad. He was trying to make software behave neatly around languages that resisted fixed rules, fixed fields and single answers.

Then Adam appeared: an extraordinarily gifted linguist

who was also a software engineer. An American who could read both Dari and Pashto, he understood the languages and the architecture. He could see where a human judgement had to be preserved and where the system could be taught to carry it.

He helped us bridge the gap.

A name is not a minor administrative field. It is often the first place a system decides whether a person can be recognised at all.

Language was not the user-interface layer. It was the first infrastructure of recognition. A system that cannot hold a name will usually fail to hold the person. The error may look small from the centre — a truncated field, a spelling variant, a failed match, a duplicated profile — but at the edge it can decide whether someone is paid, admitted, cleared, counted or lost.

The register

The hospital was still there.

Twenty years earlier, I had copied entries from its ward logs into my notebook: axe wounds, beatings, burns and injuries whose causes were unknown.

This time, someone showed me the obstetrics register.

It was enormous but soft-covered, its pages shifting under their own weight. It needed both hands. Every mother and baby had been entered in ink: the date, the delivery, the doctor's comments, the midwife's notes — and, in the margin, who came. The family members who had travelled, waited and witnessed a new person entering the world.

For many of the villages served by that hospital, this was the only record that the person existed. There might be no birth certificate, district file or database entry waiting elsewhere.

There was one line written by a midwife in one book.

I am not sure the midwife knew how precious it was. She was simply doing her work. I was the one who stood there wanting to protect it.

That book must not burn.

It must not flood, disappear during a move or be eaten slowly by damp and insects.

Manual records are often described as evidence of failure: paper waiting for proper technology to arrive. But that book was not primitive. It was the thing standing between thousands of people and administrative disappearance.

The last mile had kept its own record again.

I thought of the great books in the Peshawar bazaar, where money dealers transferred each day's transactions into permanent pages after the market closed. A dealer and a midwife appeared to belong to different worlds. In fact, they understood the same thing:

A record is a form of care.

What struck me was the arithmetic of the twenty years between my visits. Billions of dollars of aid had been programmed for Papua New Guinea, while oil, gas and gold had been drawn from the ground beneath and around these communities. Strategies, frameworks and sustainability reports had passed over the Highlands.

Yet where a human being's existence was actually established, the system still rested on one book, one line of ink and one midwife.

I felt as though I was looking backwards into the future.

The book showed what a record could protect.

It also raised a harder question: what happens when the system preserving a person's entitlement begins deciding how that person may use it?

The card and the condition

That question followed me into the Australian debate over welfare cards.

There had been fierce opposition to schemes that placed part of a person's welfare payment onto a card that could not be used for alcohol, gambling or certain cash withdrawals. Many of the communities affected had large Aboriginal populations. Critics saw the cards as discriminatory and paternalistic: another system designed by outsiders on the assumption that Aboriginal people could not be trusted to manage their own lives.

After reading *Truganini* and standing over that obstetrics register, I understood the reaction more viscerally.

A record could stop someone from disappearing.

It could also make them easier to supervise.

And yet, at almost the same time, I had been trying to design something technically similar in Syria.

At the height of the war, fears around terrorism, sanctions and banking risk had caused ordinary channels of money into almost all communities to close. One way to persuade funders and financial institutions to restart the flow was to make the money conditional: value could be released to verified people through approved shops and spent on permitted goods, while purchases such as weapons, drugs or alcohol were excluded.

There was another fear too: that the money might be converted into cash and passed to people smugglers. That was harder to classify. To an institution, it could look like diversion into irregular migration. To a family under bombardment, it might be the price of escape.

I brought my own bias to that judgement. I had never imagined devoting so much effort to solving a problem while knowing that the solution might also, however indirectly, oil the machinery of criminal trade. I had no sentimental view of the people profiting from desperation.

My concern with that machinery went back surprisingly far. Before I was even a teenager, I had somehow subscribed by post to a campaign against the proliferation of small arms and light weapons. My mother still remembers the oddity of pamphlets arriving for a child who had already decided that the spread of guns was one of the causes of the world's misery.

I guessed that criminal markets could do real harm. But I was also beginning to understand that the same transaction could look entirely different from opposite ends of it. To a smuggler, it was profit. To a family trapped by war, it might be the only route they believed remained.

In that setting, I saw the restriction as an elegant emergency fix. It could allow money to move when the alternative was no assistance at all. It gave families more dignity and choice than another box of goods selected, packed and shipped by a charity.

The objective was not to control them. It was to restore their ability to buy food, medicine and shelter from local merchants while giving frightened institutions enough assurance to release the funds.

A mechanism I understood as restoring dignity in Syria could, in Australia, evoke a long history of dignity being removed: outsiders deciding what Aboriginal people needed, where they could go, how they should live and whether they were capable of making choices for themselves.

The technology had not changed.

What had changed was my understanding of the history surrounding it.

That did not produce a simple answer. Alcohol, violence, coercion and the diversion of household money can cause real harm. Some people saw restricted cards as protection; others experienced them as humiliation.

Who had decided that a condition was necessary? Had the people subject to it helped design it? Was it imposed on everyone because of the conduct of some? Could a person challenge a decision, leave the scheme or show that the restriction was causing more harm than it prevented? Was the record being used to protect an entitlement — or merely to supervise a life more efficiently?

Amanacard was not built to police the poor. It was built to make a payment accountable without removing the human being from the centre of it. But any system capable of restricting value carries moral risk.

That power runs both ways: a card can help protect a woman's entitlement from capture; the same mechanism can become another way for outsiders to decide what she is allowed to need.

The difference is not in the plastic, the QR code or the ledger. It lies in who sets the conditions, who bears them, and whether the person subject to the system still has any power inside it.

The argument led me back to the same place as the hospital register: not to the tool itself, but to the people with the power to decide what it was for.

One of the most influential advocates of Australia's welfare card was Andrew "Twiggy" Forrest — not a welfare bureaucrat, but one of the country's best-known mining magnates. The fact that an extractive-industry figure had helped propel the idea into national policy made me look in a different direction.

If powerful companies could help shape the conditions placed around money, what responsibility did they carry for the conditions surrounding the communities from which they drew their wealth?

Beyond the meeting rooms, behind fences I had spent a career not looking over, were roads, airstrips, aircraft, power systems, communications networks and security teams — more practical capacity than any government or charity I had ever worked beside.

What would it take to point some of that capacity at a midwife's book?

9

BACK TO THE FUTURE

The men who move value

I had not fully understood where I was standing.

This was not only remote, poor and difficult territory. It was also some of the most valuable territory on the planet. For decades, oil, gas and minerals had been extracted from the earth where nearby communities still lacked basic records, reliable payments and functioning services.

Black pipelines traced the contours of the mountains, climbing through the jungle, appearing beside the roads and disappearing again into cloud above the ridgelines. Great flames rose from areas largely closed to local people. I could understand why the infrastructure appeared foreign, frightening and almost magical: an enormous hidden system entering the ground, drawing value from it and carrying that value somewhere else.

The apparatus was impossible to miss.

What remained harder to see was the return.

It seemed obvious that the companies extracting the value should help fill the gap. They had capital, machinery, communications, transport and people who knew how to operate in terrain that defeated almost everyone else. But this was the part of the system I had challenged least during my career.

I had spent decades interrogating aid agencies, banks, governments and regulators. The extractive world I had barely questioned at all.

Then, as people so often had for me in the strangest of places, someone appeared unexpectedly and opened a door.

He was a former Australian SAS man whom I met in the

middle of the Highlands jungle. He gave me a generous view of oil and gas operations from the inside — not the polished public version, but the working one: security, roads, logistics, contractors, landowners, expectations, resentment and money.

He showed me the extraordinary competence inside that world. It could move people, equipment and value through country in which aid programmes routinely surrendered to the roads. Safety could be enforced down to the last vehicle, the last pair of boots and the smallest breach of procedure. Risks were mapped. Movements were recorded. Somebody was answerable for everything that mattered to the operation.

He also knew, with an operator's precision, where the industry fell short of its own advertising.

The landscape took me back to an overland journey from Iraq into north-eastern Syria. I had driven for hours with a man who spoke no English, weaving through oilfields while war raged beyond us. For long stretches, the only movement was the pumpjacks: great metal heads bowing towards the deserted ground and lifting again, slowly and without pause, as though the war had emptied the landscape but left extraction running. Something immensely valuable was still being drawn from beneath people living by danger and improvisation above it.

I did not properly understand what I was seeing. It felt like science fiction.

Papua New Guinea held the same dissonance. Value could be extracted, secured, transported and reconciled when the system considered it important enough. Yet a woman needing safe money or a farmer needing a fair price could still disappear between the joins.

This was the inverse of the old question about paper money. Paper has little value in itself but can carry enormous value once a system recognises it. Oil, gas and gold have value before anyone writes a line about them. But intrinsic value does not automatically become recognised value for the people living above it. A resource can be measured, secured, priced globally and reconciled to the last barrel while the human claim around it remains unrecorded. The problem was never only what something was worth. It was who had authority to recognise the worth, move it, record it and answer for the consequences.

Full glory

There was a harder admission underneath this.

I had entered the aid world with a tidy moral map. We were the good intentions. Extractive companies were what good intentions protected people from.

Years earlier, I had been selected for the Young Leadership Dialogue, part of the Australian American Leadership Dialogue, a gathering that brought younger Australians into rooms with senior political, business, media and policy figures from both countries. The Washington programme included the White House. I spent much of it wondering how I had got there.

On one bus journey, Julia Gillard sat beside me and introduced herself. It was disarmingly kind and faintly ridiculous. Even before she became Australia's first female prime minister, she was not a woman who needed introducing.

Later, in an auditorium filled with figures from the Bush administration whom I knew only from newspapers and my Master's programme at Cornell University, I noticed the man beside me first by his shirt cuffs: the crispest whitest sleeves, perfectly ironed, fastened with large gold cufflinks. He looked expensive.

For small talk, I asked what he did.

He said he ran Lockheed Martin.

The girl whose pamphlets on the proliferation of small arms had once arrived by post had found herself beside the other end of the weapons industry. I could think of almost no common ground, so I asked him about corporate responsibility. The phrase was still new enough to sound like a question rather than a department, and he had an answer.

Later, over drinks, I found myself talking with Paul Wolfowitz. I had read about him, studied his policies and witnessed some of their consequences in Iraq and Afghanistan. In person, he seemed more academic and less hawkish than the figure I had constructed from a distance.

I learned that people I had encountered as symbols became more complicated at close range.

This book has been a record of tidy maps retired: the dealer who kept better books than his accusers, the pawnshops that flew in the cash, the five women who never came and the men who did.

The corporate world was perhaps the last map I had not properly tested, because it was the one my own industry most needed to be true.

Up close, I found the same coin again.

On one face was formidable operational competence.

On the other was a version of institutional theatre I recognised immediately, only better funded and more beautifully printed.

Corporate responsibility had become an industry in its own right. There were departments, consultants, targets, materiality assessments, global compacts, stakeholder maps and sustainability reports in their full glory.

The phrase itself had almost become a joke — not because companies had no responsibility, but because responsibility had been converted into a category of reporting instead of an obligation created by taking value from somebody else's ground.

The reports did not necessarily lie. That was what made them interesting. Every figure might have a source. Every photograph might have been taken. Every consultation might have occurred.

But the record answered the questions the institution had commissioned.

How many meetings were held? How many people attended? How much was spent? Which global goal did the activity support?

The harder questions were less likely to appear.

Who carried the risk? Who captured the value? Who was actually paid? Who could challenge a decision? What remained in the community when the operation, project or reporting period ended?

A company could account for a piece of machinery across thousands of kilometres while a farmer beside its road had no usable record of twenty years of production. It could monitor every person entering a secure site while the neighbouring hospital protected its births in one vulnerable book.

The capacity was already there — in people, systems, procedures, money and management attention. The revealing question was what the institution had decided was important enough to record and protect.

The aid world was not as good as its intentions, and the corporate world was not as bad as its worst headlines. Both were capable of extraordinary work. Both were also capable of producing a beautiful record of themselves while leaving the human record beneath them dangerously thin.

The responsibility of the extractive industry could not reasonably end at avoiding harm, publishing expenditure or funding a collection of community projects around the perimeter.

The wealth was not merely being generated near these communities.

It was being drawn from beneath them.

The future left running

People have often asked how I moved from English literature into political science and then into law and money.

It was not as great a leap as it might appear.

I was never a voracious reader in the sense of consuming everything. But certain books and writers changed what I could recognise. As a girl, I loved Gillian Rubinstein's *Space Demons*. Later came dystopian and existential literature: writers who showed people stranded inside systems, language losing its meaning, and machines and institutions continuing long after their purposes had become doubtful.

At University College London — founded as a secular challenge to the established order of Oxford and Cambridge — we were nevertheless sent back to the foundations. The Bible ran beneath so much of the literature that we could not understand the books without confronting it. I had not expected that in my first year.

From there, I went sideways as much as backwards: into Old English, Old Icelandic, medieval Italian and Greek philosophy. Each opened a different route into the older structures beneath the modern words I loved.

The Bible and the literature built upon it taught me to look for the distance between what institutions profess and what people do. The dystopian writers taught me to distrust machinery mistaken for progress. The existentialists taught me what it feels like to remain inside a system after its explanations have stopped making sense.

That is the context in which I want the proverb at the front of this book to be read: wisdom better than gold, judgement better than silver. In Proverbs, Wisdom is not a private consolation or an abstract principle. She is given a woman's voice and sent into public life — calling at the gates, the crossroads, the places where trade, judgement and civic life meet. The image matters here because it places wisdom exactly where systems touch lives: where value is exchanged, records are believed, and power is exercised. Wisdom does not retreat from money and power. She stands where decisions are made and asks whether they are aligned with life.

I did not really leave literature behind. Saad reminded me

of that in Syria. An unexpected chance to cross the border opened with barely twenty-four hours' notice, and my team turned it into our first meeting in five years. In that same sliver of time, Saad had a full copy of Orwell's *1984* printed and bound there, at the height of the war, and placed it in my hands when I arrived.

The visit had been improvised. The gift was exact. It was the best present I had ever received, joining the two halves of my life more neatly than I could have done myself.

I can see now that I never stopped looking for the story an institution told, the story the person inside it lived, and the gap between them.

Years later, while writing my doctorate at Oxford, a small group of us in my college became addicted to the television series *Lost*, about plane-crash survivors stranded on an island whose hidden systems and rules revealed themselves only slowly.

We would meet at improbable hours — often around one in the morning — watch for a couple of hours, then return to our rooms and our theses. It was the perfect escape from writing all day: watching another group trying to understand the rules of a place that would not explain itself.

The Highlands gave me something of that feeling. Parts of the oil and gas infrastructure had been built decades earlier. They still worked, but they creaked. Processing plants, compressors and fenced compounds sat against one of the most ancient and arresting landscapes I had ever seen. At some point the technology had been ahead of its time. Now it seemed simultaneously old and futuristic: an earlier version of the future, aged in place but never switched off.

The compounds were sealed worlds. Generators hummed through the night; floodlights held back the darkness and warning beacons blinked above steel gantries. Inside, gauges and control rooms measured everything passing through the operation, while production moved out through systems more sophisticated than almost anything available to the communities beside them.

The effect was dystopian, but not because the machines were evil.

The unease came from the separation: an advanced system moving through a human landscape without fully joining it.

I had returned to one of the first places that had overwhelmed me and found that old industrial vision still operating inside problems older than any machine.

Literature gave me a way to recognise the feeling.

It did not tell me what to do about it.

For that, I had to return to the people.

Behind the kitchens

The former SAS man continued to disturb my categories.

Nobody told me directly that he carried a mystique. I had been advised by someone else from that world to look out for him, and when I mentioned his name, people's reactions did the rest. A pause. A change in expression. The sense that I had named someone whose reputation travelled ahead of him.

It reminded me of asking about the biggest dealers in the money markets. Their names were rarely announced. They were lowered into the conversation, almost whispered, and the response told you more than any formal introduction could.

He usually ate alone or with his second-in-command, a towering Māori former police officer from New Zealand. Nobody had implied anything intimidating about the policeman. I had simply assumed, from his sheer physical presence, that he might be equally daunting.

He turned out to be a friendly giant.

More unexpectedly still, he had developed a parallel expertise in monitoring Bitcoin transactions. In the middle of the Highlands, where communications could disappear and much of the economy remained invisible on paper, I had suddenly found someone with whom I could talk about blockchain, digital value and the newest forms of financial traceability.

Security men did not always enjoy working with me. I tended to question rules whose purpose I could not see and resist systems designed mainly to demonstrate obedience. To my surprise, this did not trouble either of them. The SAS man seemed to distinguish between discipline and pointless conformity. It turned out that my tendency to buck the system quite suited him.

Then I discovered that he was a vegetarian.

The man I had imagined surviving on raw meat and intimidation was quietly navigating the woeful meat-free options in the mess. A few days later, I noticed a battered copy of Homer's *Iliad* protruding from his pack in the office. He told me he was trying to catch up on the reading he had missed during thirty years of service.

A vegetarian reading Homer in the middle of an oil and gas

operation in the Papua New Guinea Highlands was another reminder that the categories we apply to people are generally less interesting than the people themselves.

The food gave us somewhere practical to begin.

It was remarkably bad, apart from the desserts.

The desserts contained a small irony. Many were made with eggs bought locally, from chickens roaming outdoors — eggs that would probably have been marketed in Australia or Britain as premium free-range. But because they did not meet the operation's international food-safety requirements, they could not be served intact at breakfast.

They could, however, be cracked into cakes.

I began to suspect that this was why the desserts tasted so good.

Outside the fence, food was growing in abundance. Inside it, an industrial supply chain was importing frozen and packaged food across enormous distances, while some of the best local produce entered almost by accident — and disguised as sponge and apple pie.

I told the SAS man that Amanacard might help connect some of that local production to the demand inside the operation. Instead of importing so much food, perhaps we could create a reliable way to buy at least some of it from the communities nearby.

It sounded simple. It was not. A camp kitchen required predictable quantities, quality, timing and food-safety standards; smallholder farmers produced according to seasons, weather, family needs and whatever price was offered at the farm gate. Converting scattered production into a dependable commercial supply chain would require aggregation, transport, training and patient purchasing — not one ceremonial order followed by a photograph in a sustainability report.

So I began with the plate.

I traced the commodities backwards. Which vegetables could be grown locally? In what volume? At what time of year? What would need to change before a small farm could supply an institutional kitchen consistently? Which parts of the demand could be divided into achievable pieces rather than waiting for the whole supply chain to become perfect?

One of the main obstacles was good compost.

Then the friendly giant took me behind the operation to see where much of the waste went.

There I met a local man, Joseph, who had worked at the site for years and ran one of the most extraordinary operations I saw anywhere in the Highlands. The kitchens asked people to separate their waste into different bins. By the time it reached him, much of it had been mixed together again.

So he and two other men sorted through it by hand.

They separated the organic material and fed it into the machinery. Between them, they made about a metric tonne of compost a day.

I could not believe my luck.

We had been asking how smallholder farmers could improve their soil enough to grow crops reliably for the kitchens. Here was one of the essential inputs already being produced inside the fence.

And then dumped back into the jungle.

One of the men gestured towards the vegetation and said, with complete innocence, that they could not understand why the trees and plants grew so vigorously there.

The answer was lying under them by the tonne.

It was funny, but also maddening.

The formal description of waste disposal did not capture what I had seen. It could explain that the waste had been collected and processed. It did not ask the more useful question: why was a valuable agricultural input being dumped into the jungle when it could help build a local food supply chain?

That was corporate responsibility in miniature. Not another promise or community initiative, but the willingness to look closely at what the operation already possessed and place it in the hands of people who could use it.

I took the finding back. We did not get far enough to redirect the compost before the work stalled.

The answer remains behind the kitchens.

The SAS man understood the wider logic. Resource operations depend on roads remaining open and relationships remaining functional. When communities believe that money promised through landowner arrangements has not reached them, anger does not remain in a meeting room. Roads can be blocked or damaged. Operations stop. Everybody loses.

A food supply chain would not resolve those underlying disputes. But it could create a more regular and visible relationship with whole households. Farmers would have demand they could plan around. Women who grew and sold much of the food could receive value directly. The operation would obtain fresher produce and perhaps improve a diet the SAS man believed was contributing to serious health problems among the workforce.

It would be a long game.

That did not deter him. People who had already waited for decades could understand a long game, provided it was real.

When I asked him to assign members of his team to help us work with the communities, I expected some of the men who went into the field every day.

Instead, he gave me Karina, who had largely been working in administrative roles, and then Alice.

I was surprised and delighted in equal measure.

He thought they would be better.

They were.

They understood the operation but were not trapped inside its field hierarchies. They could listen without appearing to arrive with an instruction. They could sit with women, ask questions and hear things that neither foreign staff nor local men were likely to be told. Given the opportunity, they learned extraordinarily quickly.

We made modest progress. The food chain could not be transformed in a few visits, and I did not want to pretend otherwise. The most encouraging thing was how communities responded to genuine engagement and the possibility of a plan extending beyond the next reporting period.

A record of her own

The next question was how a woman selling food could remain visible after the meeting ended.

For most smallholder farmers, a cash sale leaves almost

nothing behind. The vegetables change hands. The buyer pays. The farmer returns home. Somebody farther along the chain may record the transaction, but the person who grew the food often has no record that another institution can recognise.

She may have sold produce for two decades and still appear to a bank as though she has never earned an income.

The intermediaries hold the books.

The farmers live the transactions.

The obstetrics register had shown me the danger of a whole community depending on a record held in one place. The farm gate presented the same problem in economic form. The woman existed. Her produce existed. Her work and income were real.

But the evidence belonged to somebody else — or was never created at all.

That was where the electronic handshake mattered.

When cash changed hands, the buyer and seller each confirmed the transaction using their own PIN: what had been sold, by whom, to whom and for how much.

It was deliberately simple: two people acknowledging the same event at the moment it happened. Accountability did not sit around the transaction in paperwork and later sampling. It was placed inside it.

The process also became a form of financial education.

Aid workers had told me in several countries that cash programmes were impossible because the population was illiterate. I found the assumption patronising and usually wrong. I have never worked anywhere where a household lacked someone who could read the numbers on a banknote well enough to buy the week's groceries.

People may not read a long contract. They may not use the language of banking. But they understand price, quantity, debt, change and whether someone is trying to cheat them. Numeracy, memory and repeated practice are also forms of financial literacy.

A useful system begins with those capabilities rather than using formal literacy as an excuse to deny people choice.

The technology had to make the same effort.

Our application went into Tok Pisin — the country's widely spoken creole — before it went into the field. The machinery of digital payments had to cross the language barrier, rather than expecting the customer to cross it alone. *Language* became *tokples*. *Vendor* became *man o meri bilong salim samting*. The transaction had to speak in the language in which the trade was actually happening.

There is one moment from the pilot that I will keep.

One of our team was kneeling beside a woman and her produce, showing her the record of her cabbage sales and explaining in Tok Pisin what it meant:

"Yu lukim record ya — how many cabbage you sell will be recorded here. It will stay in here as a record book for you. You hold onto it, and the balance is with you everywhere you go."

A record book of her own.

Not the hospital's book. Not the aid agency's book. Not the buyer's ledger or the intermediary's account.

Hers.

The cash was still cash. She could use it in the ways her household required. But the transaction did not vanish when the banknotes changed hands. It left behind evidence that she had grown something, sold it and been paid.

That record might one day help with another buyer, another season or an application to a bank. But its first value was more basic: the transaction became part of her history rather than merely somebody else's supply chain.

Twenty years of policy language about financial inclusion had been expressed more clearly beside a pile of cabbages.

The women already there

Alice and Karina reinforced something I had encountered throughout my working life.

The last mile of trust is very often held by women.

They may be called assistants, administrators, community mobilisers or volunteers. Their titles rarely capture their authority. They know which family has lost someone, which woman cannot travel safely, which local official can be trusted and which apparent dispute conceals a much older one.

They also know when an outsider's solution will create another burden.

Dr Pamela, a medical researcher, had spent years trying to make cervical-cancer screening work for women in Papua New Guinea. She had pioneered rapid human papillomavirus (HPV) testing at the point of care, with treatment available in the same visit where possible. The model recognised something larger than medicine: a screening appointment still requires transport, time, money and enough freedom for a woman to decide that her own health matters.

The centres were already beginning to place financial guidance alongside health services. The body and the wallet were not separate in a woman's life.

Tahina, a former professional rugby player, represented another form of practical authority. Papua New Guinea's enthusiasm for rugby league could give girls and women physical confidence, public visibility and a language of strength very different from the language of victimhood.

Reading my old notebooks made the distance of twenty years suddenly personal. I looked up people I had interviewed and discovered that many were no longer alive, among them Donna Mathew, a teacher I had grown particularly fond of, and Dale Tinol, an inspiring principal of a teachers' college.

Names that had once sat in my notebook as evidence returned as lives that had continued and ended while I was elsewhere.

I wrote to Pam and Tahina. Were the major international programmes genuinely helping? Were they reinforcing grassroots work, or taking resources and attention away from it?

It was not a rhetorical question.

I did not know. I still do not know enough.

What I knew was that I did not want our system to arrive as another external layer.

Lynda Babao, Papua New Guinea's elegant and unusually public-spirited former First Lady, told me that she was tired of hearing women in her country praised for their resilience. The word, she said, implied that they were expected to forgive, forget and adapt endlessly to whatever happened around them.

It changed how I heard a word I had used approvingly for years.

Resilience can become a compliment that functions as a sentence. It praises people for surviving conditions nobody has committed to changing.

The women I met were not abstractions of resilience. They were treating bodies, growing food, running classrooms, carrying families and finding practical routes through problems that should never have been theirs. They did not need another institution to celebrate how much they could endure.

They needed the systems around them to stop wasting their strength.

That became the test for anything we built: could it reinforce what these women were already doing without becoming one more thing they had to carry?

The test

Papua New Guinea was not a return to the past.

It was a return to the test.

The country brought together every question I had been following for twenty-five years: informal money, missing records, institutional theatre, violence, faith, extractive wealth, women's safety and the distance between what a system reports and what a person receives.

It also brought the questions closer to home.

Australia was not merely a concerned neighbour. Australian capital, expertise and institutions were woven through the resource economy. The machinery on the other side of the fence was not foreign to us.

Neither, therefore, was the responsibility.

I do not yet know how much our work will change things.

The pilot programme is young. The records are few. The obstetrics register remains a single book. A local food supply chain will require years of patient demand, not one promising conversation. The companies capable of threading pipelines up mountainsides and drawing value from beneath the jungle may still decide that protecting a midwife's ledger or a farmer's income history is somebody else's job.

But I know more than I did when I wrote *I want to scream* in my notebook.

I know that a record can keep someone visible or make them easier to control, and that technology must be judged by the power the person retains inside it.

I know that corporate responsibility means pointing the capacity to extract, secure and account for value beyond the fence — and that the people best placed to shape a solution are often the women an institution has left sitting in the office.

And I know that usefulness is not a mood.

It is a discipline.

It means returning to the places that confused you, not only to the places where your argument is easiest to prove. It means allowing the vegetarian security man with Homer in his bag, the midwife with her enormous book and the woman selling cabbages to rearrange the categories in your head.

The younger woman in the notebook wanted to scream.

The older one still does.

But now I try to build something before I do.

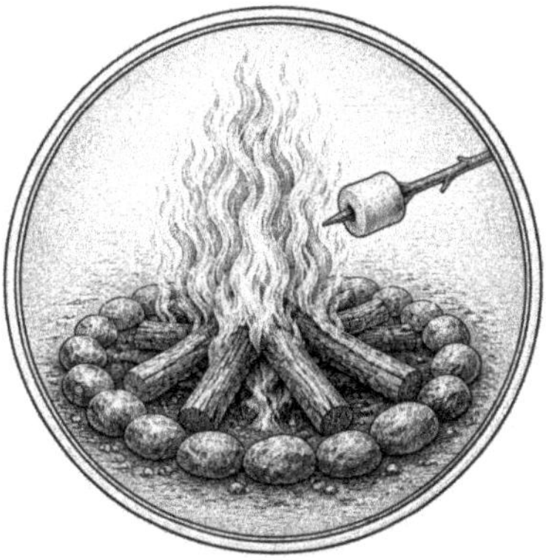

10

HUMAN JUDGEMENT

How do you conclude a book when every instinct is to push on?

There is always something urgent. A payment stalled. A system that should work and does not. A person at the far end bearing the cost of somebody else's procedure. I have spent much of my life surging towards the next problem, busy, tired, occasionally unreasonable, unable to leave alone anything I can see a way to fix. Artificial intelligence has intensified that instinct, because suddenly more of the unfinished work looks possible. The list has not shortened. It has become executable.

While trying to finish this chapter, I dreamt that my sister and I were rushing forward together. We heard no cry behind us, only a kind of silent sound, and turned to see that our father had fallen. He was hurt, but he told us to push on. I did the opposite. I grabbed him, held him so tightly that I could feel the force of it, and woke crying with my arms around nothing.

The dream was not difficult to interpret.

It is possible to become so intent on fixing what lies ahead that you fail to see what, and whom, you are leaving behind. The work can feel morally urgent because much of it is. But urgency is not a moral exemption from attention. At some point you have to turn around.

That is what writing this book has made me do.

Writing upwards

Recently, standing in a rural hospital in the Highlands of Papua New Guinea, I remembered a letter I had written more than twenty years earlier.

Back then, the health posts and clinics people actually used were overcrowded, dilapidated and packed. Nearby stood immaculate new buildings: fresh paint, modern equipment, doors chained shut. No patients. No staff. They bore signs identifying them as gifts from the Australian Government.

One clinic for people with HIV had also been placed apart from the rest, so that walking through its entrance announced to the whole community why you had come. These were not failures to build. They were failures to understand the people expected to use what had been built.

I thought surely the people who had paid for those buildings would want to know why they stood empty while patients crowded into failing facilities nearby. I found the address of the Director General of AusAID, Australia's aid agency, and wrote to him.

Years later, on the other side of the world, the Afghanistan report described in Chapter One carried a larger version of the same appeal. I had promised the Afghans we interviewed that their account would reach the parliamentary committee examining the war and be framed as a message to the British Prime Minister. At the House of Lords, the aid department's senior representative answered the messenger rather than the message.

Two continents, two aid departments and the same act of faith: send the evidence upwards and authority will act.

I knew that faith from the other side too. In Somalia, an old man once walked for days to hand me a letter asking me to feed his village. I remember his frail form, and looking directly into his eyes while I told him I could do nothing. His village was not on the list and there was barely enough food for the villages that were. The list was complete on paper and wrong in life. He had written down the need and placed it into the hands of the person who appeared to have authority. I had the paperwork, the trucks and the programme behind me. He had a letter. Neither of us had the power to change the decision.

Writing was not the opposite of doing. It was one form of doing — a way to carry an observation beyond the room in which it might otherwise disappear. But it ended by handing the next move to somebody else.

Eventually I began building as well as writing. Not because the record had ceased to matter, but because I had stopped assuming it would act by itself. A report could describe the gap. A working system could be placed inside it and tested under load.

I needed both. Writing preserved what action was liable to outrun. Building revealed whether an idea could survive contact with the world.

The record

My family has always known that I think better when I can see the thing in front of me. My sister has spent much of her career in very flash jobs with access to magnificent printers, an advantage I have exploited shamelessly for reports, proposals and every book I have attempted to write. A document changes for me when it becomes physical. I spread pages across tables and floors. I see repetitions, gaps and false joins that were invisible on a screen. I write in the margins, draw arrows, move whole sections around. The printed word does not replace thought. It gives thought somewhere to stand.

That same distinction — between an aid to thought and a substitute for it — returned in a conversation with Saad, the man who assembled the original Syrian team with me and has spent nearly ten years helping turn impossible field problems into workable ones. I had been showing him how a large language model such as Claude could take on some of the repetitive work we had dragged behind us for years: comparing records, reorganising material, compiling first drafts and transcribing data between different forms. Not so that it could think for him, but so that he had more time for the judgement only he could bring.

He understood immediately that it represented a new stage in how we could work. Then, characteristically, he challenged me to consider the consequence: what might handing part of our work to a machine do to how we think? He asked whether I had a moment. With Saad, that usually means a profound observation is about to arrive.

He told me about Socrates. I had just shown a Syrian field operator how to hand ten years of administrative drudgery to an artificial intelligence, and he answered with Plato. It was both very Saad and a considerably better philosophical objection than most of those offered in the podcasts I listen to on my daily walks.

The walks themselves had begun as obedience to a word. For years I kept a Post-it note from Martha de Richemont stuck to my screen in London. It said only: MOVE. Martha, who shared my birthday and lived to 104, left an instruction that now seems to run against the mythology of the moment: intelligence detached from bodies, judgement lifted into systems, speed mistaken for life. MOVE sent me the other way — back into a body, a day, a route through trees, and the knowledge that decisions are still made by people who must live with what follows. Those walks gave what Saad had said room to work on me.

In Plato's *Phaedrus*, Socrates recounts a warning about the invention of writing: people would come to rely on external marks rather than their own memory, and might acquire the appearance of understanding without understanding at all. Written words were especially suspect because they could not answer a question or defend themselves when misunderstood. Saad's conclusion was that Socrates would turn in his grave at Claude.

He had a point. A fluent machine can produce the appearance of understanding at a scale Socrates could not have imagined. Claude can even do the one thing a page cannot: answer back. That may make the warning more important, not less. An answer is not the same as accountability, and fluency is not judgement.

But I cannot agree that external memory necessarily weakens thought. The written word has strengthened mine all my life. It allows experience to survive the mood and exhaustion of the day. It lets one person inspect what another saw, and someone years later notice what both missed. It does not have to end a conversation. At its best, it makes a longer and more crowded conversation possible.

I was struck by this again while listening to an interview about *Careless People*, Sarah Wynn-Williams's account of seven years inside Facebook, now Meta, and the decisions of its senior leadership. Meta had obtained an arbitration order restricting her from promoting the book or publicly criticising the company, so the interview was built around what she was legally constrained from saying. Yet the book itself remained in circulation. It could travel into rooms she could not enter and keep saying what she had already set down. That is not proof that a book is sufficient. It is proof that a record can outlast the conditions imposed on the person who made it.

A book cannot answer back. It can, however, place people who never met, and may never agree, around the same table.

The campfire

It turns out I have always been preoccupied by how people make decisions. One of my sister's favourite stories about me is that, before I was ten and still happily ensconced in Melbourne, I would regularly emerge from my bedroom announcing that I had written a book — usually less a book than a fully illustrated guide to something urgent, such as how to survive a fire and what, precisely, *stop*, *drop* and *roll* required of the reader.

So perhaps it was not such a leap that, years later, I became immediately absorbed by a drawing on the back of a paper placemat in Geneva. I had met two jolly Australian human-rights lawyers, Phil Lynch and Ben Schokman, beside a baggage carousel, where I had correctly predicted that their luggage would not arrive. The missing bags gave three strangers something to connect over while the empty belt went round, and we ended up having dinner that night.

I told them what I had been grappling with in the field: difficult decisions made under pressure by people who each held a different part of the picture, but with no simple way to show afterwards what had been considered, whose judgement had counted, or why one course had prevailed. Phil and Ben listened in the way good lawyers do, asking questions that forced me to separate instinct from reason. Then one of them turned over the placemat and began to scribble.

By the end of dinner, the problem I had carried in my head had been distilled into a simple form: the assumptions made visible, the Red Cross principles stated, the reasons preserved. They mapped it onto the legal test of proportionality — it was a beautifully ironic crossover: using high-level legal doctrine to anchor a boots-on-the-ground crisis workflow. The form was never intended to make the decision. Its purpose was to slow a group down just enough for different kinds of knowledge to enter the same judgement.

My friend Oz and his friend Can later turned the placemat into software, coding the tool we called Decision-Navigator. I watched Oz smoke into the night and type furiously in a language I had no way of accessing, while the thing in my head appeared on a screen. But software was never the idea. It was an enabler. It was easier to carry into a field site than bundles of paper and pens, easier to use around a campfire, easier to preserve and revisit after the people involved had moved on.

The tool travelled into settings and decisions I could not have imagined — thirty-two countries in all. Teams found different uses for it, each revealing something new about the decision in front of them. Across those applications, one pattern kept returning. The best judgement did not come from the most senior person, the cleverest specialist or the person who spoke with the greatest confidence.

It came when people who saw the problem differently were made to think together: different disciplines, ages, nationalities, genders, levels of authority and kinds of experience. The lawyer noticed one risk, the logistician another, the local colleague the thing both outsiders had missed. The person responsible for security saw the danger of acting. The person living with the programme saw the danger of delay. None held the whole decision. Together they produced a richer decision set than any of them could have produced alone.

One dilemma still tests the point. In Myanmar, after Cyclone Nargis, the fastest way to move relief at scale was on the government's own military assets — the same military the affected population had every reason to fear. Speed argued one way. The security specialist saw what association with those assets could cost us later; the local colleague saw what accepting them would signal to people still deciding whether to trust us at all. The form did not make the choice, and it could not make it painless. What it did was force every one of those observations into the same decision, and keep the reasons on the record — where they could be inspected, and defended, long after the water receded.

What was decided is not for these pages. The register protects the decision as carefully as it preserves the reasoning behind it — that is the point: private where it should be, accountable where it matters. But there is no doubt that mapping the dilemma against the proportionality test helped. This was not an exceptional day. It was one of the daily decisions needed, and the form gave it the discipline its weight deserved.

Difference was not friction to be managed; it was the method. The tool could not force people to be honest or brave, and it was never going to banish hierarchy from the room. What it did instead was far more practical: it stripped the loudest voice of its power to erase quieter, critical observations, while inviting senior leaders to co-own and defend the eventual decision — making them truly accountable for the tough choices. By preserving what was considered, what was set aside and why, the tool made the reasoning more legible without pretending the decision had been made by a machine.

I have since seen the same principle everywhere. It was there in Kabul when money dealers taught an Australian researcher what financial officials had missed. It was there in Geneva when lawyers gave operational form to field instincts. It was there when Saad assembled seven Syrians who could hardly have been less alike and whose differences became the strength of the team. It was there at my kitchen table when an exiled Afghan security official and I brought two entirely different bodies of memory to three thousand names. It has been there in the relationship between Maryam and the men who support her work, between my father's attention to the human gesture and my fixation on the ledger, and between Oz's code and the field problem I could describe but not build.

The best ideas in my career have appeared in the space between unlikely minds.

That is why the campfire matters. The strongest group is not the one with the most impressive people or the greatest quantity of information. It is the one with enough difference to expose its own blind spots, enough trust for disagreement to survive, and enough proximity to keep the people who will live with the decision visible while everyone else decides.

The newest partner

Artificial intelligence has now taken a place at the campfire.

Working within the safeguards required to protect personal and commercially sensitive information, I have been able to give Claude a breadth of context no single colleague could reasonably have been asked to absorb: field notes, research, operating decisions, source code, discarded designs and unfinished ideas accumulated across a quarter of a century. In minutes, it can retrieve, compare, question and synthesise material from across that record, placing decisions made years apart alongside one another. It has helped me document the platform we have built as fully as I always wanted — already more than eight hundred pages and still growing. It has helped me read my own company's source code and rediscover decisions I had almost forgotten making.

That does not make it the source of judgement. It makes collaboration with my own past possible.

The distinction matters, particularly for anyone who manages people. For bounded, checkable work, the machine can already be better: faster, more consistent and less prone to omission. Point it towards a feature or function and it can produce clean documentation across hundreds of pages, then revise it repeatedly without embarrassment, defensiveness or fatigue. It does not require the challenge to be wrapped in pleasantries; directness can improve the answer. That does not mean it is always right. It means correction does not become a relationship problem. For anyone trying to run a bootstrapped company through a war, that is not an abstract advantage.

That advantage changes what is scarce. Execution is becoming abundant. Attention is not. The more work a machine can set in motion, the more carefully the people directing it must decide what it may do alone, what must remain open to inspection and where a person must still answer.

But the answer cannot be to romanticise human inefficiency. Tea matters because judgement happens while it is being drunk, not because slowness is virtuous. A machine can document a verification procedure. It cannot sit with a merchant until he decides what he trusts you enough to show; notice the person who has not spoken; understand what an introduction has cost; or bear the consequence when the conclusion is wrong. People still have to decide and remain answerable, especially when somebody else carries the cost of error. Their irreplaceable contribution is not inconsistency. It is presence, context, courage, care and judgement.

And there are forms of understanding no archive can provide. My father came with me on this improbable journey for ten of its years and has known me for all of mine. When I was too ill to stand, he stepped into workshops from Pakistan to Mozambique and carried the work forward. A machine can hold more of my records in view than he ever could. It cannot know what to do because I am not myself that day. Context is not the same as having been there, and neither is the same as love.

Oz recently reminded me of another part of this. Over a beer, long after the placemat, he told me that my father and I had changed how he worked. At first, he said, our need to know everything had felt like micromanagement. Then he understood that we were not asking in order to catch him out. We wanted to know so we could help think the problem through with him. He said it had shaped the way he manages people and now the way he works with AI. The original collaboration had travelled into another person's practice and then into a technology none of us could have imagined when he first coded Decision-Navigator.

That is the kind of loop I want to close. Not the loop in which I am proved right, but the one in which an idea becomes useful to somebody else.

The misjudgement

The letter to Canberra and the report in London were not mistakes because I wrote them. The mistake was the expectation I attached to them.

I believed that if we made the evidence clearer, the reasons more visible and the consequences harder to hide, reasonable institutions would make different choices. Decision-Navigator, the safer-corridor proposals and Amanacard were arguments made executable. They showed that another way was possible.

Often, the institutions still did not move.

The problem was not always ignorance, or even misunderstanding. Existing arrangements worked tolerably well for those near the centre. Procedure protected the institution; mandate protected the contractor; risk policy protected the bank; the funded intermediary protected the donor. The person at the far end could remain unseen without anyone feeling individually responsible for the outcome.

A system that makes outcomes visible also makes responsibility visible. I had underestimated how unwelcome that could be.

My misjudgement was not that software could decide; I had never believed that. It was that institutions with the power to act would necessarily want better-supported human judgement once it became available. I assumed evidence was scarce. Often appetite was scarcer.

The people at the other end were never confused about the value of what we built: the nurse receiving her wages, the family receiving the remittance, the merchant whose honesty could finally be seen, the woman whose work could no longer be erased from the record.

The work had not been misplaced. I had mistaken the institution for the customer.

Human judgement

So what have I been studying all these years?

I began with money, because money leaves trails even

when people would prefer it not to. Then I wrote about trust, because no payment system operates without somebody accepting somebody else's promise. But underneath both was judgement: whom to listen to, what to notice, which risks matter, whose account counts, when to act, when to refuse, and how to remain answerable for the decision afterwards.

Good judgement is not a private possession. It is a social achievement.

It depends on who is invited into the room, whose experience is treated as knowledge, whether dissent survives hierarchy, whether the assumptions are visible, whether the reasons are recorded and whether the people who will live with the consequences remain present in the decision. Technology can widen the circle, preserve the record and help us see more. It cannot decide what we owe one another.

I remember the old shock of *Total Recall*, the 1990 film with Arnold Schwarzenegger and Sharon Stone: not a machine revealing itself through a glitch, exactly, but something more disturbing. The woman occupying the most intimate role in the protagonist's life turns out to be part of *The System* managing him. The danger now is not that simulated intimacy will reveal its machinery, but that it may work well enough to be trusted.

The final boundary is relationship. Money can create trust between strangers. Records can preserve recognition across distance. Language can order a world. But none of them is safe when separated from people who remain answerable for what the words, records and payments do. Wisdom, in the old sense, is not superior processing. It is judgement in right relationship with reality: close enough to see, humble enough to listen, brave enough to decide and responsible enough to answer for the outcome.

This is the part I want to carry forward. The future will not be improved by replacing fallible people with machines that make fallibility harder to see, or by systems that invite trust and intimacy while their accountability lies elsewhere — and perhaps, eventually, nowhere at all. Nor will it be improved by romanticising human intuition. We need better-formed judgement: more diverse minds, more complete records, more honest disagreement and a much shorter distance between the decision and the person who bears its cost.

I finished these pages in Provence, near Arles and Saint-Rémy, surrounded by a landscape Van Gogh had made alive to me when I was still a schoolgirl: cypresses, wheatfields, hot blue sky, all charged by the force of his looking. He was another child of 30 March, which had thrilled me when I first read his letters to Theo. But the deeper force was his insistence that seeing is not passive. It can be urgent, costly and years ahead of those ready to receive it.

The dream about my father returned me to that distance. He told us to push on, because that is what fathers say when they do not want to become the reason their daughters stop. But the right response was not to obey immediately. It was to turn around, hold him and notice that he had fallen.

Perhaps that is how to conclude a book too.

You stop surging forward for long enough to gather what made the work possible: the people, the unfinished ideas, the letters that went unanswered, the paper records, the jokes, the warnings you filed away and the observations that only make sense twenty years later. You do not preserve them because the past was better. You preserve them because the next judgement will be poorer without them.

I cannot send a reply to the old man in Somalia. I do not know his name or what became of his village. But his letter remains at the centre of this book.

I could do nothing for him that day. I have spent much of my working life trying to build systems that can write back: the name is on the list; the list reflects the person; the money arrives addressed; the reason for the decision can be inspected; the person at the far end does not disappear inside somebody else's procedure. The banks call where he stood the last mile. It was always the first.

A conclusion is not where I claim that work is finished. It is where I decide what must not be lost.

So I am putting the record down in the middle of the campfire — not as a verdict, and not as a machine's answer, but as material for the next judgement.

The rest depends, as it always has, on who gathers around it — and whether the man with the letter is still visible when they decide.

www.ingramcontent.com/pod-product-compliance
Lightning Source LLC
La Vergne TN
LVHW051919051126
851858LV00023B/315/J
* 9 7 8 1 7 6 4 8 4 2 6 3 1 *